Praise for *Last Ape Standing*

"[An] engaging account . . . shed[s] a fascinating light on our evolutionary success." —*The New Yorker*

"Walter takes an antic delight in the triumphal adaptations and terrifying near misses of human evolution . . . *Last Ape Standing* makes for a lively journey." —*The New York Times Book Review*

"Chip Walter has made himself indispensable to audiences craving the latest information about our evolutionary past. No one writes about early man, evolutionary dead ends or our pre-human rivals better than Chip Walter. If all science books were this witty and well-written, everyone would be a nerd." —*Pittsburgh Post-Gazette*

"Chip Walter's *Last Ape Standing* is provocative, insightful, and engaging; a rare trifecta among science books. Nearly every page offers something that will surprise or intrigue you."
—Ray Kurzweil, inventor, futurist, and author of
How to Create a Mind: The Secret of Human Thought Revealed

"[An] engrossing, up-to-date account of human evolution."
Kirkus Reviews

"I read *Last Ape Standing* while sitting, then I jumped up and cheered. It's that good!" **—William Shatner**

"The saga of human evolution is far from a straight line from ape to angel, with all but one of many species going extinct. Chip Walter's thoroughly enjoyable new book considers the evolutionary and social forces that crafted us, modern humans, and presents an intriguing scenario of why Homo sapiens is the Last Ape Standing."
—Donald Johanson, discoverer of Lucy and founding
director of the Institute of Human Origins
at Arizona State University

"[A] captivating and informative field trip through man's paleontological past . . . an exceptionally well-written overview of man's evolutionary history as well as an accessible guide to the underappreciated field of paleoanthropology." —*Booklist*

"This book has a way of making you feel magnificently insignificant and at the same time an essential, vital part of the chain of human evolution. Just when you thought you were fully evolved as a human . . . think again. Mind blowing stuff!" —**Michael Keaton**

LAST
APE
STANDING

*The Seven-Million-Year Story of
How and Why We Survived*

CHIP WALTER

BLOOMSBURY
NEW YORK · LONDON · NEW DELHI · SYDNEY

Black and white image credits. Images 1, 2, 3, 7, 10, 11, and insets within 5, 6, and 12: Frank Harris. Image 4 inset of *Paranthropus aethiopicus*: Sergio Pérez. Image 9: based on an image provided by the National Institute of Health, 2010. Image 8: based on a drawing by T. L. Lentz, originally published in *Primitive Nervous Systems* (New Haven: Yale University Press, 1968).

Published by Bloomsbury USA, New York

All papers used by Bloomsbury USA are natural, recyclable products made from wood grown in well-managed forests. The manufacturing processes conform to the environmental regulations of the country of origin.

LIBRARY OF CONGRESS CATALOGING-IN-PUBLICATION DATA HAS BEEN APPLIED FOR.
ISBN: 978-0-8027-1756-6 (hardcover)

First published by Walker & Co. in 2013
This paperback edition published in 2014

Paperback ISBN: 978-1-62040-521-5

1   3   5   7   9   10   8   6   4   2

Typeset by Westchester Book Group
Printed and bound in the U.S.A. by Thomson-Shore Inc., Dexter, Michigan

For Cyn. My compass and my Gibraltar.

CONTENTS

AUTHOR'S NOTE

Despite its academic-sounding name, a good deal of brawling often goes on within the field of paleoanthropology. That it explores the deep past and counts on bits of ossified bone grudgingly revealed or scraped out of the earth doesn't help the inexactness of the science, or the disagreements it generates. Although all researchers in the field work hard to bring the objectivity of the scientific method to their work, its nature involves a lot of guesswork. So while one scientist or group of scientists may think that the unearthed fossils of a particular creature demand that it be classified as a new species, others might feel just as strongly that it is simply a new example of a species that has already been discovered. Some scientists find good reason to have created the classification *Homo antecessor,* for example. Others, just as reputable, and just as thorough in their thinking, argue no such species ever existed.

No one really knows. The evidence is too sparse and too random. We are making up these names as a convenient way of organizing the chaos of discovery over the past 180 years. It's not as though the creatures themselves went by the nomenclatures we have made up. Nor can we comprehend what we don't know. We can never say if we have discovered the fragmented evidence of 80 percent of our direct ancestors and cousin human species, or 1 percent.

Too often, being human, we may give the impression we understand more than we do, or that we have just about figured it all out. We haven't, as you will see. One of the reasons this book is relevant is because the human family tree, or more precisely, our very limited view of it, has changed so much in just the past five years.

Advances in genetics, innovations in radiocarbon dating, together with plain old scientific creativity and elbow grease have greatly improved our guesswork and helped flesh out the discoveries we have made. There would be no hope, for example, of having even the remotest idea that a wisdom tooth and the end of a pinkie finger found in a Siberian cave three years ago belonged to an entirely new species of human (scientists call them Denisovans) with whom we and Neanderthals may share a common ancestor. This paltry evidence even revealed we mated with them! Nor would we have learned that billions of humans (including, very possibly, you) have Neanderthal blood running in their veins. But we now know these astonishing things are true, even as they have turned assumptions once taken as gospel entirely on their heads.

Still, despite these advances and the exciting discoveries they have made possible, the illumination of our past is a little like trying to find a set of car keys in the Sahara with a flashlight.

I bring this up now to clarify a point: we don't know exactly how many other human species have evolved over the past 7 million years—27 or 2700. We likely never will. But I have tried to arrive at an arguable and acceptable number that makes the larger point that, despite the disagreements that take place within the field, the story of how we came to be is a good deal more intriguing and complicated than we thought even a few years ago. And that makes the story even better.

INTRODUCTION

Over the past 180 years we have so far managed to stumble across, unearth, and otherwise bring to light evidence that twenty-seven separate human (*hominin*, to use the up-to-date scientific term)[1] species have evolved on planet Earth. As you may have noticed, twenty-six of them are now no longer with us, done in by their environments, predators, disease, or the unfortunate shortcomings of their DNA. The lone survivor is a peculiar, and peculiarly successful, upright walking primate that calls itself, a little self-importantly, *Homo sapiens sapiens*, the wise, wise one. In most circles, we simply call them you and me.[2]

Of all the varieties of humans who have come and struggled and wandered and evolved, why are we the only one still standing? Couldn't more than one version have survived and coexisted with us in a world as big as ours? Lions and tigers, panthers and mountain lions, coexist. Gorillas, orangutans, bonobos, and chimpanzees do as well (if barely). Two kinds of elephants and multiple versions of dolphins, finches, sharks, bears, and beetles inhabit the planet. Yet only one kind of human. Why?

Most of us believe that we alone survived because we never had any company in the first place. According to this thinking, we evolved serially, from a single procession of gifted ancestors, each replacing the previous model once evolution had gotten around to getting it right. And so we moved step by step (Aristotle called this the "Great Chain of Being"), improving from the primal and incompetent to the modern and perfectly honed. Given that view, it would be impossible for us to have any contemporaries. Who else could have existed, except our direct, and extinguished, antecedents? And where else could it all lead, except to us, the final, perfect result?

This turns out to be entirely wrong. Of the twenty-seven human species that have so far been discovered (and we are likely yet to discover far more), a considerable number of them lived side by side. They competed, sometimes they may have mated, more than once one variety likely did others in either by murdering them outright or simply outcompeting them for limited resources. We are still scrounging and scraping for the answers, but learning more all the time.

If we hope to place our arrival on the scene in any sort of perspective, it's a good idea to remember that every species on Earth, and every species that has ever lived on Earth (by some estimates thirty billion of them), enjoyed a long and checkered past. Each came from somewhere quite different from where it ended up, usually by a circuitous, and startling, route. It's difficult to imagine, for example, that the blue whales that now swim the world's oceans, great leviathan submarines that they are, were once furry, hoofed animals that roamed the plains south of the Himalayas fifty-three million years ago. Or that chickens and ostriches are the improbable descendants of dinosaurs. Or that horses were once small-brained little mammals not much taller than your average cat with a long tail to match. And the Pekinese lapdogs that grace the couches of so many homes around the world can trace their beginnings to the lithe and lethal gray wolves of northern Eurasia.

The point is, behind every living thing lies a captivating tale of how the forces of nature and chance transformed them, step by genetic step, into the creatures they are today. We are no exception. You and I have also come to the present by a circuitous and startling route, and once we were quite different from the way we are now.

Theories about our ancestry have been amended often because new discoveries about how we came into existence keep emerging; several times, in fact, while this book was being written. But however it played out in the details, we know this: for every variety of human that has come and gone, including those we think we have identified as our direct predecessors, it has been a punishing seven million years. Survival has always been a full-time job, and the slipperiest of goals. (It still is for most humans on the planet. More than four billion people— nearly two thirds of the human race—subsist each day on less than two dollars). But luckily, for you and me at least, while evolution's turbulent dance rendered the last line of non–*Homo sapiens* DNA obsolete eleven thousand years ago, it allowed ours to continue until finally, of all the many human species that had once existed, we found ourselves the last ape standing.

Not that we should rejoice at the demise of those others. We owe a lot to the fellow humans that came before us—hairier, taller, shorter, angrier, clumsier, faster, stronger, dumber, tougher—because every one of us is the happy recipient of all the more successful traits that our ancestors acquired in their brawl to keep themselves up and running. If today we were to make the face-to-face acquaintance of an *Australopithecus afarensis* or *Homo ergaster* or *Paranthropus robustus*, what would we see? Intelligence, fear, and curiosity is my guess, for starters. And they would see the same in us because we truly are kindred spirits.

This has ensured that many of the deft genetic strategies that made those now departed human species once possible still remain encoded in the DNA that you and I have hauled with us out of the womb and blithely carry around each day into our personal worlds. The millions of these creatures who came and strove and passed through the incomprehensibly long epochs between their time on earth and the here and now are, after all, us, or at least some of them were. We are a marvelous and intriguing amalgamation of those seven million years of evolutionary experimentation and tomfoolery. If not for the hard planks of human behavior those others long ago laid down, we would be naught.

So you can thank the lines of primates that in eons past found their way into Africa's savannas, then to Arabia and the steppes of Asia, the mountain forests of Europe, and the damp archipelagoes of the Pacific, for genetic innovations like your big toe, your ample brain, language, music, and opposable thumbs, not to mention a good deal of your personal likes and dislikes, fascinations, sexual proclivities, desires, temper, charm and good looks. Human love, greed, heroics, envy, and violence all trace the threads of their origins back to the deoxyribonucleic acid of the humans who came before us.

Some might wonder what sense it makes to rummage through the leavings of the past seven million years to try to piece together the story of our peculiar emergence. The payoff is that it's the best way to understand why we do the startling, astonishing, sometimes sublime and sometimes horrible things we do. We owe it to ourselves to unravel the riddles of our evolution because we, more than any other animal, *can*. If we don't, we stand no chance of comprehending who we really are as individuals or as a species. And only by understanding can we hope to solve the problems we create. To not understand how we came into this universe damns us to remain mystified by our mistakes, and unable to build a future that is not simply human, but also humane.

This in itself, however, fails to answer the nagging question of why our particular branch in the human family managed to find its way to the present when so many others were shown the evolutionary door. Plenty of other human species had a good run; many considerably longer than ours. Some were bigger, some were stronger and faster, some even had heftier brains, yet none of those traits was good enough to see them through to the present. What events, what forces, twists, and evolutionary legerdemain made creatures like you and me and the other seven billion of us who currently walk the earth possible?

Somehow only we, against all odds, and despite the brutal, capricious ways of nature, managed to survive . . . so far.

Why?

Our story begins once upon a time, a long time ago . . .

Chip Walter
Pittsburgh, Pennsylvania
June 2012

CHAPTER ONE

THE BATTLE FOR SURVIVAL

DNA neither cares nor knows. DNA just is.
And we dance to its music.
—Richard Dawkins

T HE UNIVERSE HOUSES, by our best count, one hundred billion galaxies. In one sector, a not very remarkable galaxy in the
shape of a Frisbee with a bulge at the center spins pinwheel style
through the immense void. One hundred billion suns reside in the
Milky Way, each annihilating—at varying levels of violence—
uncounted trillions of hydrogen molecules into helium. Along the
edge of this disk, where the star clusters begin to thin out, sits the sun
that we wake up to every morning. By some cosmic calculus that science has yet to decipher, the planet we call home came to rest at just
the right distance away from that star, and with just the right makeup
of atmosphere, gravity, and chemistry to have made an immense variety of living things possible.

Our universe has been around roughly fifteen billion years; our sun,
perhaps six; Earth has made four billion circuits around its mother star,
and the life upon it has been ardently evolving during 3.8 billion of
those journeys, give or take a few hundred million years. For the vast
majority of that time anything alive on Earth was no larger than a single
cell. Had we been around to see this life, we would have missed it since
it is invisible to the naked eye. But of course, if it hadn't come first, we,
and everything else alive on Earth, would have been out of the question.

No matter how hard we try, we cannot begin to fathom the changes, iterations, and alterations our planet has undergone since it first came to rest, collided and molten, in its current orbit. Our minds aren't built to handle numbers that large or experiences that alien. And in this book we won't try. Instead we will focus on a hiccup in that Brobdingnagian history, but a crucial hiccup, especially from your particular point of view—the past seven million years. Because that is when the first humans came into existence.

Compared with the other smaller, rocky planets in our solar system, Earth has always been especially capricious. During its lifetime it has been hot and molten, usually wet, sometimes cold, at other times seared. Occasionally large sectors have been encased in ice; other epochs have seen it covered with gargantuan swamps and impenetrable rain forests inhabited by insects larger than a Saint Bernard. Deserts have advanced and retreated like marauding armies, while whole seas and oceans have spilled this way and that. Its landmasses have a way of wandering around like pancakes on a buttered griddle, so that the global map we find familiar today was entirely different a billion years ago and has spent most of the time in between resolutely rearranging itself, often with interesting results for the creatures trying to survive its geologic schemes.

Charles Darwin and Alfred Russel Wallace figured out by the mid-nineteenth century that these incessant alterations explain why Earth has spawned so many varieties of life. Random revisions in the DNA of living things, together with erratic modifications in environments, have a way of causing, over time, whole new and astonishing forms of life to emerge. Darwin, when after years of thought and hand-wringing finally got around to writing *On the Origin of Species*, called this "descent by means of natural selection," which was to say that creatures change randomly (by mechanisms unknown to him; he and everyone else in 1859 were ignorant of mutating genes or spiraling ladders of DNA) and either endure in their environment, or not. If the mutations that shift their traits help the organism to survive, he surmised, then it breeds more offspring and the species continues, passing the new traits on. If not—and this has been the case with 99.99 percent of all of the life that has ever evolved—then the life form is, as scientists like to say, "selected out." Darwin figured that different environments favored different mutations, and over time—incomprehensibly long periods of time—different organisms diverged like the branches of a tree, each shoot putting increasing distance between the others around it until, eventually, you find your self with varieties of life emerging as different as a paramecium and Marilyn Monroe.

So, though we may have started out with single prokaryotic cells and stromatolitic mats of algae nearly four thousand million years ago, eventually the world found itself brimming, in stages, with lungfish, slime mold, velociraptors, dodo birds, salmon and clown fish, dung beetles, ichthyosaurs, angler fish, army ants, bighorn sheep, and— after almost all of the time that ever was had passed—human beings— peculiarly complicated animals, with big brains, keen eyes, gregarious natures, nimble hands, and more self-aware than any other creature to have ever come down the evolutionary pike.

Of the twenty-seven species of humans that we have so far found that once walked Earth, ours is the most favored line for the simple reason that, so far at least, we have avoided the genetic trash bin. Given evolution's haphazard ways, we might just as well have ended up a blowhole-breathing water mammal, a round-eyed, nocturnal marsupial, or a sticky-tongue anteater obsessed with poking its wobbling proboscis into the nearest nest of formicidae.* We could even have *become* the formicidae for that matter.

Or we might have become extinct.

But as it turns out—and lucky for us—we emerged from the jungles of Africa, came to stand upright, clustered ourselves into close-knit packs, gave up our front paws for hands, grew thumbs, took up re-formed, meat-eating diets, developed tools, and, in a remarkably short time—as evolutionary events go—rearranged the world right down to its molecules and right up to its climate. Today we are even manipulating the DNA that makes us possible in the first place—a case of evolution evolving new ways to evolve. (Think about that for a moment.)

We did not pop out of the jungles of Africa in our current form like Athena from the head of Zeus, big-brained, tool-laden, and ready for modern life. We came in stages, part of a vast and jumbled experiment largely driven by our home planet's fickle ways. Between six and seven million years ago Africa's rain forests began to shrink slowly. Earth was a different place from today, but not radically so. If there

---

*Ants.

were time-traveling, Earth-orbiting satellites to provide a global view of our planet then, it would look pretty much the same as the one we see on the Weather Channel today. India would be in place mostly, though still slowly plowing into Asia, creating the Himalayas. Australia would be roughly located where we find it now. The Mediterranean would be a touch larger. Partially submerged, the boot of Italy would not look quite so bootlike, and the Bosporus strait along with sectors of the Middle East would be inundated, though soon enough the closure of the strait at Gibraltar would transform the Mediterranean into a vast plain of salt flats, marshes, and brackish lakes.

These geologic alterations were unfolding because the planet was warming up, thinning its ice caps and making land scarcer and Earth more watery. Ironically, the world was becoming much like the one scientists now speculate global warming is creating. In looking back at our origins, it seems we are catching a glimpse of our future.

Climate, however, is complex. Weather systems veer and fluctuate. Tectonic plates beneath the Indian Ocean were shifting and sloshing whole seas. As the planet generally warmed, some parts of the world became wetter, and more tropical, while others grew drier. Among these were northeast and north-central Africa, where grasslands were gradually transforming themselves into desert, and rain forests were breaking up into semiwooded savannas. Here, a new kind of primate was evolving, probably several. Primates that were not, purely speaking, any longer creatures of the jungle.

Scientists peg the emergence of Earth's first human about seven million years ago largely because around that time the fossil evidence, sparse as it is, points to a primate splitting from the last common ancestor we shared with today's chimpanzees. There is no precise method for fixing dates of these kinds. Paleoanthropology, with its reliance on the chance discovery of ancient bones and the sediments in which they lie, is replete with perplexity and, as sciences go, is far from exact.

In fact, the likelihood of any ancient bone even becoming a fossil is so vanishingly small that it's just short of miraculous any discoveries are made at all. If you hoped, for example, that some part of you might be discovered fully fossilized in the distant future, there would be no chance of its happening unless you dropped dead in a layer of soft sediment that takes an impression of your body, or into some place lacking the oxygen that so enthusiastically decomposes every molecule of us when we bite the dust. A peat bog or shallow, muddy river would be a good place.

From there you would have to hope that the tectonic shenanigans of the planet, lashings of wind and water and climate, the shifting courses of rivers, or the encroachments of deserts or glaciers wouldn't toss or shuttle your bones from their resting place to some location less hospitable to your preservation. Assuming they didn't, then at least some of the solid parts of your remains would have to be replaced molecule by molecule with other dissolved solids that leave behind a stone replica of your formerly carbon-made skeleton. Then finally, if all of this happens just so, you must count on the wind or rain or the instinct of an exceedingly lucky paleoanthropologist to reveal what is left of you to him or to her.

The chances of your being preserved in this way are, by some estimates, one in a billion. The likelihood of this small part of you then actually being found is so small, it can't accurately be calculated. Add to this that many of our earliest ancestors met their fate in forests or jungles where decomposition happens rapidly and without leaving a trace, and you can see why the fossil record we rely upon to unlock our origins is not only tiny, but serendipitously skewed. At best we have been left with random clues that provide only the sketchiest images of the deep past. In fact, whole lines of primeval relatives were almost certainly long ago obliterated and now lie beyond discovery.

We do have tools other than fossils that can help divulge our ancestry. The science of genetics is still fledgling, but it provides ways to explore the past by providing a kind of clock that allows scientists to estimate when certain branches of our family tree made off in different directions. (See the sidebar "Genetic Time Machines," page 76.) Yet the best genetic evidence is currently so foggy that it places the time we and chimpanzees shared a common ancestor somewhere between four and seven million years ago, rather a loose estimate. So neither the fossil record nor genetic science can provide anything very detailed about the precise time of our emergence.

Still, we have to start somewhere. It sometimes shocks people to learn that at least twenty-six other human species once lived on earth. It further shocks them that many of them lived side by side. The point is there was not, as we often think, an orderly march of ape-men that led from chimp to you and me.

One reason science has tentatively settled on seven million years as the birth date of the human species is that the oldest fossil that might reasonably lay claim to being human was found in Chad at various times between July 2001 to March 2002 (he was unearthed piece-meal). His discoverer, a student named Ahounta Djimdoumalbaye, called him *Sahelanthropus tchadensis*—Sahel man, after the part of Africa south of the Sahara where he was found. Not much remained of this particular primate—a skull, four lower-jaw fragments, and a few teeth, but because the fossils indicated his head was positioned much like ours is, in line with his torso rather than at a forty-five-degree angle like a knuckle-walking gorilla, some paleoanthropologists speculate he (or she) walked upright. They see this as a reason to consider him (or her) an early human. All we know for certain is that *tchadensis* was either one of the last ancestors humans shared with other great forest apes or was one of the first humans to have evolved. Or *tchadensis* might be an evolutionary dead end. The best we can say is, the bones left behind were found in sediments that tell us *tchadensis* walked the earth about seven million years ago, and so that is where we shall begin.[1]

When compared with the billions of years it has taken to make a universe or its suns and planets, seven million years may appear minute, but to those of us who aren't stars, comets, oceans or mountain ranges it remains a very, very long time. We are used to measuring time in hours and days, months and years, perhaps generations when forced to push the envelope. Epochs and eons bend the mind and are as incomprehensible as light-year-measured galactic distances or quantum calculations computed in qubits.

To help wrap our minds around these numbers, imagine that we could squeeze the seven million years that have passed between the arrival of *Sahelanthropus tchadensis* and the present into a single year's calendar, and then plot the arrival—and in some cases the departure—of every known human species from January to December. Let's call this the Human Evolutionary Calendar or HEC. If we look at it this way, *tchadensis* arrives January 1. Lucy, the famous upright walking member of a line of savanna apes known as *Australopithecus afarensis*, who lived about 3.3 million years ago, appears July 15. Neanderthals

The Human Evolutionary Calendar - Evolution in a Year

January ————— March ————————————— June ————————————— September ————————————— → December

Sahelanthropus tchadensis
7,000,000 - 6,000,000 yrs.

Ororin tugenensis
6,100,000 - 5,800,000 yrs.

Ardipithecus kadabba
5,750,000 - 5,200,000 yrs.

Ardipithecus ramidus
3,200,000 - 4,300,000 yrs.

Australopithecus anamensis
4,200,000 - 3,900,000 yrs.

Australopithecus afarensis
3,500,000 - 2,900,000 yrs.

Australopithecus africanus
3,800,000 - 3,000,000 yrs.

Kenyanthropus platyops
3,500,000 - 3,200,000 yrs.

Paranthropus aethiopicus
2,650,000 - 2,300,000 yrs.

Australopithecus garhi
2,750,000 - 2,400,000 yrs.

Australopithecus sediba
2,000,000 - 1,750,000 yrs.

Homo rudolfensis
1,900,000 - 1,750,000 yrs.

Homo Ergaster
1,900,000 - 1,3,000,000 yrs.

Homo erectus
1,800,000 - 250,000 yrs.

Homo habilis
2,350,000 - 1,450,000 yrs.

Homo georgicus
1,800,000 - 1,3000,000 yrs.

Paranthropus bcisei
2,275,000 - 1,250,000 yrs.

Homo rhodesiensis
300,000 - 125,000 yrs.

Homo pekinensis
700,000 - 500,000 yrs.

Homo heidelbergensis
700,000 - 200,000 yrs.

Homo antecessor
1,000,000 - 700,000 yrs.

Homo neanderthalensis
200,000 - 28,000 yrs.

Denisova hominins
200,000 - 30,000 yrs.

Paranthropus robustus/crassidens
1,750,000 - 1,200,000 yrs.

Red Deer Cave People
? - 11,000 yrs.

Homo sapiens sapiens
200,000 - 0 yrs.

Homo floresiensis
100,000 - 13,000 yrs.

don't show up until near Thanksgiving, November 19, and we *Homo sapiens sapiens* finally reveal ourselves near the winter solstice, December 21, a little more than a week before the end of the year.

Looking at this timeline, you can't help but conclude the human species seems to have gotten off to a slow start, at least based on the current sketchy evidence.* Following *tchadensis* nothing at all happens for more than a million years, then a creature researchers call *Orrorin tugenensis* (Millennium Man) finally appears just before the spring equinox—on March 8. Like *tchadensis*, *tugenensis* didn't leave much for us to inspect—two jaw fragments and three molars. Later finds turned up a right arm bone and a small piece of thigh—altogether enough information for paleontologists to conclude that *Orrorin* was almost certainly human, and lived about 5.65 to 6.2 million years ago, mostly in wet grasslands and fairly thick forests that eventually became the Tungen Hills of modern Kenya. Thus the name *tugenensis*. Whether he walked upright all the time or even part of the time is debated, but if he spent his days between grasslands and jungle, he may have done a bit of both, walking on all fours in the forest and upright now and again in and among the trees and grasslands he called home.

As we move into spring not one, but three new and indisputably human species arrive. On March 18 two emerge from the mists of time: *Ardipithecus ramidus* and *Ardipithecus kadabba*; then on May 20, *Australopithecus anamensis*. These were all distinct species, yet all three bear a stronger resemblance to today's chimpanzees than to us, and all three probably walked upright sometimes, and at other times on all fours.

By summer in the HEC, signs emerge that the human experiment was gathering momentum. Multiple species begin to appear and overlap. Recalling their names is a little like trying to follow the characters in a Russian novel, but bear with me. (We can thank the brilliant zoologist Carl Linnaeus for the long and respected tradition of assigning elongated, Latin names to all living things.) In mid-October, *Paranthropus robustus* (sometimes known as *Paranthropus crassidens*) arrives. Then on July 4, *Kenyanthropus platyops*; ten days later, *Australopithecus afarensis* (Lucy); and then in August, *Paranthropus aethiopicus* and *Australopithecus garhi* join the ranks of humans that have walked the planet.

---

*Many more human species may have existed at this time, but the farther back in time you go, the more likely those creatures lived in rain forests, and the less likely conditions were optimal for creating fossils.

These creatures, each of whom found their way in and out of time and the plains and forests of Africa, arose and departed subject to the cantankerous whims of evolution. When we compress time this way, it's easy to forget that some of these species lived for hundreds of thousands of years. All of them were intelligent, with brains that ranged from the size of today's chimps, 350 cubic centimeters (cc) to as large as 500 cc, still a quarter to a third the size of the brain modern humans carry around, but enormous and enormously complex when compared with those of most other mammals.

Something strange and intriguing was afoot across Africa's sprawling lands. Like an Olympian god, the continent's changing climate was forcing the emergence of multiple kinds of humans, all of them descended from jungle primates similar to those that still live in Africa's rain forests today (albeit in ever-dwindling numbers). In time the selective pressure that different environments exerted coupled with random genetic changes resulted in new varieties of humans that emerged all over the continent.

Aethiopicus arose along the banks of Lake Turkana in Kenya and the Omo River basin of Ethiopia. Lucy and her kind roamed as far north as the Gulf of Aden and as far south as the ancient volcanoes of modern Tanzania, while *Australopithecus africanus* lived thousands of miles south, not far from Johannesburg, South Africa. A later addition to the human family, dubbed *Australopithecus sediba*, was recently discovered in South Africa as well. The partial skeletons of a young boy, and an adult female, who lived between 1.78 and 1.95 million years ago (between mid and late October), were scraped from the dust.

Depending on where they lived, all of these species dealt with surroundings that ranged from densely wooded and fairly wet, to dry, open grassland. As Africa's jungles retreated toward the center of the continent, troops of apes must have been left scattered over hundreds of thousands of square miles to adapt or die. They had no tools, only the randomly provided equipment their genes conferred upon them, all better fit for life in the jungle than the environment they now faced. Where once the rain forest provided them with ready supplies of fruits and berries that delivered plenty of energy and nutrients, they now found themselves dealing with savannas where less food was spread out over larger areas, inhabited by growing numbers of predators each exceedingly focused on making meals of them. Life was, in the immortal words of Thomas Hobbes, "poor, nasty, brutish, and short." Everything was more dangerous, and staying alive demanded more energy, mobility, toughness, and cunning.

Wherever they lived and however they survived, all the hominin primates that emerged during the summer in the Human Evolutionary Calendar were players in a grand, African experiment now three million years in the making. The world was testing them, harshly, and the forces of evolution were remorselessly molding them into a new kind of ape.

While the random forces of evolution endowed each with different genetic attributes that helped them all survive, every one of them seems to have developed one predominant trait: for the first time in the incomprehensibly long story of evolution on earth, species had emerged that walked upright on their hind legs. Because we do this so effortlessly every day, it may escape you how exceptionally strange this mode of transportation was four million years ago among living mammals, or any other animal for that matter. But strange it was. Yet precisely because it was so peculiar, it set in motion a string of evolutionary events that eventually enabled you and me to come into existence.

We are so steeped in technology, so used to controlling our environment, that we forget that the only way the vast majority of living things can hope to survive in a changing world is to come by the right genetic mutation at just the perfect time, something that happens entirely by accident. Serendipity is the enemy and the ally of every species. It might provide you with the claws needed to bring down prey or the speed required to escape another's claws. Or it might not, in which case you are doomed to be "selected out," unfit for your new habitat and relegated to the genetic landfill. For living creatures of all kinds, and that included our ancestors living during the balmy summer months of the HEC, there are no evolutionary shortcuts, no quick technological fixes, no ways to take charge and change the rules of the game with an invention.

But sometimes you get lucky.

If you stand back and look at the sprawling landscape of life's evolution on Earth, it's easier to pick out big trends, and that can help clarify a mystery or two. For example, when similar creatures find themselves in similar situations, they sometimes develop nearly identical traits, but by entirely separate evolutionary paths. Take seals, dolphins, and whales.

All of them were former land mammals, but each developed fins. They didn't, and couldn't, inherit these traits from one another because they were distinct species that evolved independently. But because living in water seems to favor creatures that grow fins of some kind, each shares this trait. Scientists call this convergent evolution.

Something like this seems to have happened with several lines of savanna apes beginning around four million years ago. While they all descended from jungle cousins who walked on all fours, many forsook knuckle-walking. And it makes evolutionary sense that they did. In the jungle, food is never far away: plenty of low-hanging fruit. Wild gorillas, for example, travel only about a third of a mile on average every day, sometimes only a few hundred feet. Everything they could ask for is close by.

On the savanna, though, life was profoundly different. Beneath the hot equatorial sun, temperatures would often have risen into triple digits (Fahrenheit). Food was scarce and rarely at hand. So while walking upright in the thick underbrush of a tropical rain forest would have done nothing to improve your chances of living a longer life in the jungle—in fact, it might shorten it—perambulating on your hind legs in open grasslands provided several advantages. You gained better visual command of your world, which is useful if you are on the daily menu of ancient jackals, hyenas, and the lion-size, saber-toothed cats called megantereons. Traveling on two feet is also more efficient than scrambling along on four. Studies reveal that knuckle-walking chimpanzees burn up to 35 percent more energy than we humans do as we stroll blithely down the street. Making your way around the broad, hot grasslands of the Pleistocene epoch on your knuckles and hind legs looking for food, watching out for predators and taking care of your young, would have been slow, tiring, and ultimately deadly. That is presumably why upright walking became the preferred form of navigation for all savanna apes no matter what locality they called home. Those that failed to come by this trait were wiped out.

Precisely how ancient humans such as Lucy, *aethiopicus*, and *Australopithecus africanus* eventually pulled off the physical tricks necessary to stand upright remains a mystery, but they did, and one reason they did is thanks to a genetic trait common to all apes—a big toe.[2]

For some time zoologists have known that early in gestation the big toe of gorillas, chimps, and bonobos is not bent, thumblike, but is straight, similar to ours. But as they continue their development the

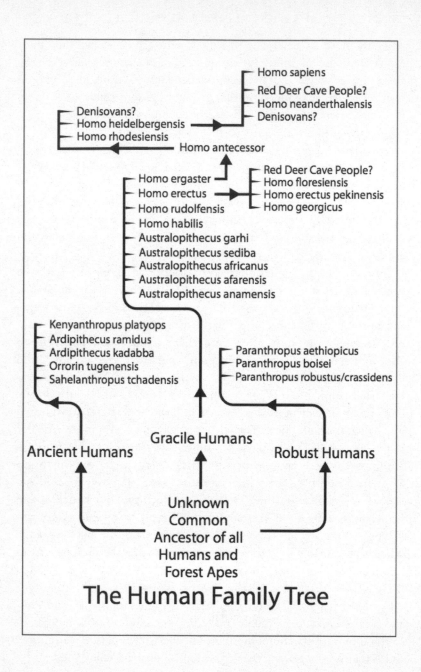

The Human Family Tree

big toe departs from the other four so that by the time they are born it has become thumblike, making it easy for their feet to grasp, stand on, or hang from limbs. But what if one of the descendants of a jungle ape found itself living in sparse forests and open savannas? And what if one of those apes was born with a big toe that never became thumblike, but instead remained straight, a freak genetic deformity?

Deformities, autoimmune diseases, even mental illnesses, are often the result of genetic mutations. Somehow a gene is reshuffled, a hormone misfires, or a genetic switch is delayed. Even DNA makes mistakes. In fact, evolution depends upon it. We humans are sometimes born with an extra digit, webbed fingers, shortened legs. But one creature's deformity can become another's salvation. Every living thing on Earth is, one way or another, an amalgamation of genetic gaffes.

Imagine, then, that some primates were born "deformed" with a straighter toe preserved from their time in the womb rather than a more opposable one like all other normal primates. What sort of life could such a creature look forward to? In the jungle, not a promising one. Unable to effortlessly grasp tree branches like his fellow apes, he would struggle mightily to keep up with the troop, die swiftly, and his genetic predilection for straight toes would bite the dust with him.

But in a partially wooded savanna, where the grasslands were expansive and the forest broken and less dense, an ape with a straight big toe would be lucky indeed. That deformity would enable him to stand and walk upright. Without a straight big toe our current brand of walking would be impossible. With every step we take, our big toes support 30 percent of our weight, and they make the upright running, jumping, and rapid shifts in direction we excel at possible. Other complicated anatomical changes had to have taken place before our style of two-footed walking came into existence, and eons passed before those modifications were completed, but the transformation arguably began with a straight big toe. Such an odd foot would have made any savanna ape better at standing upright and able to walk for longer periods on his hind legs. In time he would have found himself endowed with a birth defect that would eventually prove a lifesaver.

Moving in this odd, vertical way didn't mean that ancient humans entirely forsook swinging in trees or walking on their knuckles. Lucy, who provided paleoanthropologists with one of the most complete skeletons of an early ancestor, appears to have been a hybrid, clearly capable of walking on her knuckles if she felt like it, and outfitted with

Australopithecus afarensis

shoulders and arms that were nicely adapted to swinging through and climbing trees. Yet, the architecture of her pelvis, the tilt of her head, and the shape of her foot tell us that upright walking was her preferred way of getting around; so preferred that her footsteps in wet mud or sand would have looked almost exactly like yours or mine.

Walking upright was one evolutionary trait our predecessors shared as they marched resolutely to the present, but not the only one. Another was in play that would also have enormous ramifications: their brains were growing larger. Not immensely larger, but the difference is measurable. While a chimp's brain is about 350 cc, these grassland primates' brains ran from 450 cc to 500 cc, a 25 to 40 percent increase.

The big question is why. The traditional scientific answer to this question is that a bigger brain is a better brain, so evolution's forces tended to favor smarter animals. That is true enough, but it still doesn't explain the mechanism that was causing the growth. What was forcing the issue? Why were larger brains evolving at all? Strange as it may seem, starvation might be the answer.

When an animal is having a chronically difficult time filling its belly, something intriguing happens in its body at a molecular level. Aging slows down, and cells don't die as quickly as they do when food is available. Contrary to what you might think, a cell's health in this situation doesn't deteriorate. It improves. The body, sensing deprivation, seems to call all hands on deck, husbands its energy, and prepares for the worst. In a sense each cell grows tougher and more wary. This is thanks largely to a class of proteins called sirtuins, which some scientists suspect reduce the rate of cell growth.[3]

Numerous studies show that reducing the normal diets of creatures as different as fruit flies, mice, rats, and dogs by 35 to 40 percent will increase life span as much as 30 percent. (Scientists can't ethically perform these sorts of experiments on humans, but all indications are that the same holds true for us.) When food is scarce, fertility also drops and animals mate less frequently, an additional way of slowing down the cycle of life. While the deprivation makes life horrible for the creatures enduring it, from an evolutionary point of view it carries with it the quality of pure genius. Nutritional penury not only extends the life of an animal, but fewer offspring improves the chances of the whole species remaining in the evolutionary sweepstakes. Fewer offspring also places less stress on already overburdened food resources. The whole process of living decides, it seems, to bide its time until the storm passes. Cell growth on every level slows except for one key and remarkable exception: brain-cell growth increases.

There the cells last longer, *and* they begin to make new versions of themselves faster, or at least the neurotrophins generated by the hypo-thalamus, which are the precursors of new brain cells, do. Not only that, other experiments show that food deprivation increases an appetite-stimulating peptide called ghrelin, which enables synapses to transform themselves by some molecular magic into cortical neurons. You could say the body and the brain strike a bargain. To compensate for the aggressive growth of new neurons, the rest of the anatomy fasts, stretching scarce nutritional resources that it then redirects to the brain. Or put another way, the body slows down aging and accelerates intelligence. This means that 3.5 million years ago, by the time Lucy and her contemporaries were desperately scavenging at the margins of an unpredictable land, the chronic deprivation they were facing was accelerating the growth of their brains.[4]

So our ancestors had two assets going for them. Upright walking made them more mobile and efficient, able to cover more ground and better equipped to evade the predators that were evolving along with them. Their larger brains meanwhile made them more capable of adapting to dangerous situations on the fly, more adept scavengers, and better at collaborating successfully with one another. All good in these strange and dangerous environments. That they survived despite their desperate circumstances proves that the combination of the two adaptations was succeeding. But there was now a new challenge: the two trends were on a collision course and bound, in time, to make it impossible to survive. Something had to give.

CHAPTER TWO

THE INVENTION OF CHILDHOOD (OR WHY IT HURTS TO HAVE A BABY)

My mother groaned, my father wept,
into the dangerous world I leapt.
—William Blake

The human birth canal inlet is larger transversely than it is anteroposteriorly
(front to back) because bipedal efficiency favors a shorter anteroposterior
distance between a line that passes through both hip joints and the
sacrum . . . This size relationship, along with a twisted birth canal shape,
makes human parturition mechanically difficult.
—Robert G. Franciscus
"When Did the Modern Pattern of Childbirth Arise?,"
Proceedings of the National Academy of Sciences

TWO AND A half million years ago, around the end of August in the Human Evolutionary Calendar, primates like *Kenyanthropus platyops*, *Australopithecus afarensis*, and *Australopithecus africanus* begin to disappear from the fossil record. They may not actually have disappeared, but the evidence of them does. Either way, their evolutionary run was apparently nearing an end. However, with their disappearance a new, rich wave of human species began to crest and break on Africa's broad and windy plains. In the space of one million years, nine new varieties of humans emerged. Stepping back and looking at the aggregated remains scientists have labored to pick out of the hills, valleys, and ancient lake beds of Africa like so many needles from an incomprehensibly huge haystack, these species give you the impression that the savanna apes that had been struggling so long to survive were finally getting the hang of living in their new environment, fanning out in more directions, deepening the peculiar evolutionary experiment we call humanity.

And it *was* an experiment; make no mistake, because not all branches of the human family were evolving along the same lines. More precisely, species were striding down two distinctly different roads—one that included smaller, slimmer, so-called gracile apes, and another that embraced bigger, thicker humans with large jaws and teeth, known in the world of paleoanthropology as robust apes. Each approach had its advantages. But in the long run, only one would succeed.

The members of the robust branch of the human family first showed their simian faces in late August among the flooded grassland along the Omo River in southern Ethiopia and the western shores of Lake Turkana in northern Kenya. Scientists call this specimen *Paranthropus aethiopicus*, and he is perplexing because he combines so many contradictory characteristics. His bone structure seems to say that he more often than not walked on all fours among the elephants, saber-toothed cats, and hyenas with which he passed his days. Yet he lived in wet, open grasslands munching on tubers and roots with his big, flat teeth and ample jaws, rather than in wooded areas where you might think knuckle-walking would make more sense. Despite his chimplike anatomy and relatively small brain (no more than 450 cc in adulthood), he may have been the first to pull off the astounding trick of fashioning the first stone tools, preceding even the famous feats of "Handyman" (*Homo habilis*), who followed him and is generally

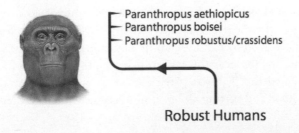

Paranthropus aethiopicus
Paranthropus boisei
Paranthropus robustus/crassidens

Robust Humans

considered the inventor of the first Neolithic technology. (Scientists are debating anew who should get credit for this remarkable advance.)

Whatever *aethiopicus* accomplished, more like him were to follow. Later in the calendar year—the middle of October—two other *Paranthropus* species, *boisei* and *robustus* (also known as *crassidens* in the ever-changing argot of paleoanthropology), arrived, also generously jawed, large headed, and big of tooth, like *aethiopicus*.

Paranthropus humans represent an evolutionary "strategy" that modified the behaviors of jungle apes, but didn't leap dangerously far from them. Of the two routes down which evolution was walking earth's humans, this was the safer, more conservative one. Like their predecessors in the rain forests, troops of robust apes roamed from location to location, gathering what food they could find in the thinning forests, bush, and grasslands where they lived. Because of the sorts of foods they ate, *Paranthropus* possessed heads that sported thick, sagittal crests like the ones you see on the silver-backed gorillas at your local zoo, though they were more chimp-size than gorilla-size. The crests are a stout, jagged line of bone that runs from the top of the forehead to the back of the neck like the metal rim of a medieval helmet. Anchored to these were thick ropes of muscle that ran to their massive jaws and dense necks so that the broad, square rows of teeth in their mouths could crush the cement-hard shells of the nuts they consumed, pulverize bark and seedy berries, crunch the exoskeletons of large insects, or masticate the bones of an unfortunate small animal they might have been lucky enough to snatch up.

Beneath these crests the brains of *boisei* and *crassidens* had expanded roughly a third during the four million years that had passed since the first human emerged from Africa's rain forests. They were undoubtedly resourceful and even more socially bonded than the apes from which they had descended, mostly thanks to the menace that surrounded them. Danger breeds reliance and cooperation. Day-to-day living would have been unimaginably harsh: a life of slow migration, eating to gather the strength to move forward and moving forward to gather more food to eat. Despite its hardships, however, this was by no means an unsuccessful evolutionary path. By current accounts, *boisei* roamed the plains of Africa for a million years, foraging the foods at hand and getting along, if not famously, then well enough. If we measure success by how long species survive, we *Homo sapiens*, amount to little more than rookies still wet behind the ears. We have been in the game of life a scant two hundred thousand years. *Boisei* held sway on the Horn of Africa five times as long before exiting the gene pool. If we become this lucky, we will someday be dating our letters July 12, 802013.

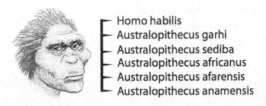

Homo habilis
Australopithecus garhi
Australopithecus sediba
Australopithecus africanus
Australopithecus afarensis
Australopithecus anamensis

The other path plotted by the combinations of genes, environment and random chance was the one taken by members of a branch of the human family paleoanthropologists like to call gracile. This includes *Australopithecus garhi*, a creature who, along with *aethiopicus*, made his debut on the Horn of Africa about 2.5 to 3 million years ago. There is some slim evidence that *garhi* may also have fashioned simple stone tools, but as in the case of *aethiopicus* that's a controversial and unresolved theory. At best, *garhi* may have used crude stone hammers to break open bones to get at the marrow inside, or sharp flint to scrape and hack meat away from a bone left behind by larger predators. But even these uses of rock mark a colossal technological advance.

About 1.9 million years ago another gracile human, dubbed *Homo rudolfensis*, appeared along the shores of Lake Rudolph, now known as Lake Turkana, a long body of water that runs in the shape of an index finger from southern Ethiopia into the western heart of Kenya. *Homo habilis* and *Homo ergaster* soon followed, both slim and light-boned, both also passing their time in East Africa.

In 1991 scientists scrounging among rocks near Dmanisi, Georgia, west of the Caspian Sea, unearthed the remains of still another species of gracile human from this epoch—*Homo georgicus*. While he remained simian in his looks, his face was flatter, a step closer to ours. Like *Homo habilis*, *georgicus* was a lean toolmaker, but with a considerably more advanced case of wanderlust. He lived in a river valley more than twenty-five hundred miles north of the grasslands where

Homo habilis passed his days. He may be an indicator that other species, so far unknown, also strode beyond the borderlands of the Dark Continent, settling who knows where, still awaiting discovery.

Although all of these species came upon the world clustered, like a posse, information about the majority of them is sketchy. Drawing any conclusions about them is a little like drawing conclusions about a long-lost family relative who headed off to the merchant marine or the French Foreign Legion. The best we have in most cases is a few battered bones that offer scant insights into the creatures' lifestyles or appearance. *Georgicus*, for example, has seen fit to provide us with three skulls—one with jaws attached, one with a solitary jawbone, and one missing its jaws altogether. Nor did they leave anything much in the way of teeth, let alone whole limbs or vertebrae. *Homo rudolfensis* has bequeathed a similarly ungenerous array of jaws and skulls, and a scattering of other fragments that may not belong to the species. What we know of *ergaster* (the Workman) is based on a bundle of six or so skulls, jawbones, and a few teeth, several of which don't much resemble one another, creating some lively academic brawls about exactly which species is which.

The stinginess of these creatures makes them mysterious, even among our ancestors, humans who have steadfastly held the cards of their pasts close to their primeval vests. Of all these slender primates, however, one has been a little less secretive—*Homo habilis*, otherwise famously known as Handyman, long thought to be our direct ancestor and the first toolmaking primate. We have been able to infer a little more about the life of *habilis* only because we have been lucky enough to have stumbled across more parts of his body than his other contemporaries—several skulls, a hand bone complete with fingers, and multiple leg and foot bones that can't conclusively be connected with the skulls, but at least provide some clues about the creature's size and gait. Together the evidence tells us that *habilis*, though slight in stature, walked upright all the time and possessed considerably larger brains than the first ancient humans, as spacious as 950 cc, depending on which skull you inspect. The shapes of their heads and jaws indicate that unlike their robust cousins, they didn't care much for nuts, bark, and berries, but had developed an appetite for meat, and the protein it provided, which may account for their larger brains. (See sidebar "Big Guts vs. Big Brains" p. 21.) Nor did they sport great sagittal crests, or

Big Guts vs. Big Brains

Cows, as we all learned in grade school, have four stomachs. They do because it requires a lot of work to extract enough nutrients from grass to transform it into beef and milk. The same was true of our early savanna-roaming ancestors, at least some of them. Subsisting on a diet of nuts, roots, thistles, berries, and other plants required long intestines and strong stomachs if they hoped to squeeze enough nutrients from them to stay alive.

As the climate changed in Africa and the savannas became broader and drier, the old jungle ways of gathering low-hanging fruit from nearby trees and not moving very far from day to day simply didn't work. Fruit and foliage became increasingly rare, and three humans had to cover more distance to gather it, which required still more energy. Ultimately that was not a sustainable survival strategy.

But if you could get your hands on some meat! Then you were instantly rewarded with much more nutritional bang for your hunting-and-gathering buck. That is precisely what the robust lines of savanna humans did. But this choice paid an additional, unexpected dividend. A diet of meat of any kind (even dining on termites and small rodents) made larger brains possible, and less cowlike intestinal tracts necessary. This is something paleoanthropologist Leslie Aiello dubbed the Expensive Tissue Hypothesis when she first came up with the idea in the early 1990s. What this meant, and what fossil finds reveal, is that as our ancestors began to consume more meat, their bodies could redirect the energy those complex intestinal tracts demanded to the business of constructing larger brains. It was a close question two million years ago which approach might work best. Both experiments were tried, and for hundreds of thousands of years both worked. Ultimately, though, larger brains turned out to be a more effective survival tool than longer intestines, something the fossil record bears out. While australopithecines and the robust members of the human family were relatively small brained, often not much more cerebrally endowed than a chimpanzee, *Homo ergaster*'s brain size ballooned to 900 cc or so. After a run of more than one million years, the last of the robust humans finally made their exit 1.2 million years ago.

This evolutionary path had other ramifications as well. We aren't as strong as our primate cousins—chimps, gorillas, orangutans—for example. We seem to have exchanged brawn for brains. Richard Wrangham has argued that mastering fire and cooking made meat and other foods of all kinds easier to digest, increasing the protein we could consume and reducing the need for longer intestinal tracts even further. In time bigger brains delivered better weapons, and more strategic ways of hunting. And that likely led to bigger game, more meat, more protein, more cerebral horsepower. The result? Over the past two million years, the brains of the gracile line of humans nearly doubled their size.

huge, square teeth made for grinding. Their teeth were better at tearing. Chances are they hunted small game in packs, not unlike the way chimpanzees sometimes do. And they helped themselves to savanna carrion and whatever other more adept and deadly predators left behind in the way of their prey's remains.

The scattered fossils of both of these sides of the human family tell us that evolution was putting a series of unstated questions on the table 1.5 million years ago: Which approach is best? Gracile or robust? A steady diet of tubers, nuts, and berries? Or a sparse, starvation diet of scavenged carrion along with whatever else could be scraped from nature's table. A serviceable brain with a cast-iron stomach, or a great brain with a simpler, less sturdy digestive system?

If you were a betting primate, you couldn't be blamed for putting your money on the robust branch of the family. At the time, they looked to be winning the battle. They were strong and durable and had adapted the jungle ways of their antecedents to the savanna exceedingly well, meandering through flooded grasslands and clusters of forests, sometimes upright, sometimes on all fours, consuming, if not jungle fruit like their more gorilla-like ancestors, then the next-closest foods that give the term *high fiber* a whole new meaning.

Their stomachs had to be large and their intestines long to digest these foods, and their eating would itself have required liberal funds of energy. In some ways they were consuming so they would have the energy to consume. According to one theory, this explains why their brains did not grow as large and as fast as their gracile cousins'. Hard-working stomachs can't afford to redirect energy to cerebral growth. But their arrested development may have saved them and been the secret to the success of their million-year run.

Gracile apes on the other hand looked less likely to succeed. They were smarter—given their increased brain size they had to be—but their diet was unpredictable. They used less energy because they walked upright all the time, but they had to make do with whatever else their smaller, less sturdy stomachs could handle. The robust approach was stable. The gracile approach was risky.

Sometimes, however, risk pays off. A high-stakes wager placed on *Paranthropus* would have paid nothing, yet against all odds, several underdog gracile species remained in the hunt. Good news for us because it was from one of these lines that you and I descended. Still, trouble loomed. Just as it appeared gracile apes were succeeding, finally brainy and efficient enough to outfox the rough treatment their savanna environment was dishing out, the self-same adaptations that were saving them—an upright gait and bigger brains—were also aligning to become the agents of their doom.[1]

Striding on two legs efficiently—not waddling the way a chimp or gorilla does when it walks upright—requires, among other adaptations, a fundamental rearrangement of pelvic architecture. An upright stride narrows the hips, and for females, narrowing the hips narrows the birth canal, and a slimmer birth canal makes for increasingly snug trips for newborns out of the womb. Despite the many advantages that upright walking delivered, it creates problems when one is simultaneously evolving bigger brains and larger heads, which was precisely what our gracile ancestors were up to. Yet, since both adaptations were working, what could be done? Each was an evolutionary blessing, yet both were on a collision course. Something would have to give.

Lucky for us, the forces of evolution worked out an exceedingly clever solution: gracile humans began to bring their children into the world early. We know this because you and I, being extreme versions of gracile apes, are the living, breathing proof. If you, for example, were to be born as physically mature and as ready to take on the world as a gorilla newborn, you would have to spend not nine months in the womb, but twenty, and that would clearly be unacceptable to your mother. Or, looked at from a gorilla's point of view, we humans are born eleven months "premature." We do not reach full term, which makes us fetal apes. Of course if we didn't make our departure from the womb ahead of schedule, we wouldn't be born at all because our heads, after nearly two years in the womb, would be far too large too make an exit. We would be, literally, unbearable.

It's impossible to overstate the colossal impact this turn of events had on our evolution, but it requires some context to fully appreciate what it means. Our habit of being born early is part of a larger, stranger phenomenon that scientists call *neoteny*, a term that covers a lot of evolutionary sins at the same time it explains so much of what makes us the unique, even bizarre creatures we are.

The dictionary defines *neoteny* as "the retention of juvenile features in the adult animal." The term comes from two Greek words, *neos*, meaning "new" (in the sense of "juvenile"), and *teinein*, meaning to "extend." In our case it meant that our ancestors, rather remarkably, passed along to us a way to stretch youth farther into life. The question is, why, and how, did it happen?

When faced with resolute obstacles, evolution—always in the service of survival—has a marvelous way of selecting astonishingly diverse solutions cooked up entirely by random chance. This is how the planet has found itself with the unearthly-looking aye-aye of Madagascar, Borneo's clownish proboscis monkey, the squashed and unappetizing blobfish of Tasmania, and the rapier-nosed narwhals of the arctic seas. It also helps explain the bizarre mating rituals of porcupines, and male anglerfish, not to mention the torturous eating habits of ichneumon wasps. Each of these creatures is a living testament to the marvelous, if accidental, creativity natural selection conjures, again and again. But as remarkable as these evolutionary banks and turns have been, neoteny can count itself as one of the strangest, and we *Homo sapiens* are by far the most dramatic and extreme example.[2]

The term *neoteny* was coined by Julius Kollmann—a groundbreaking German embryologist and a contemporary of Charles Darwin's. Kollman had nothing like human beings in mind when he created the term. He conceived it to describe the retention of larval features in the Mexican axolotl (*Ambystoma mexicanum*), and other species of salamanders like the mud puppy (*Necturus maculosus*) and the olm (*Proteus*), all of which refuse in their lives to fully grow up and out of their larval stage, even in their adulthood. They mature normally and sexually, but all within the body of their youth. This would be a little bit like a two-year-old boy behaving in every way like a fully grown, sexually mature twenty-five-year-old. In humans, neoteny isn't quite that pronounced (probably a good thing), but it is nevertheless remarkable, and remarkably odd, if you are willing to circle around and look at it fresh.

The idea of neoteny predates even Darwin and was explored as far back as 1836, when Étienne Geoffroy Saint-Hilaire, a French scientific prodigy and compatriot of Napoléon's, first pointed out how astonishing it was that the young orangutans that had recently arrived from Asia at the Paris zoo resembled "the childlike and gracious features of man."

In the twentieth century a handful of other scientists and evolutionary thinkers adopted Kollmann's term and Geoffroy's sentiments when they began applying the idea of neoteny to humans, observing that infant apes bore a striking resemblance to adult humans especially in the shapes of their faces and heads. Naturally this raised a few questions: Was this simply a coincidence? Why would we resemble baby apes? And did this have anything to do with our own evolution?

A professor of anatomy in Amsterdam named Louis Bolk became nearly obsessed with those questions. Between 1915 and 1929 he penned six detailed scientific papers and one entire pamphlet on the subject with the ambitious title *Das Problem der Menschwerdung* ("On the Problem of Anthropogenesis"). He argued that a surprisingly high number of human physical traits "have all one feature in common, they are fetal conditions [seen in apes] that have become permanent [in adult humans]."[3]

In one paper Bolk even enumerated twenty-five specific fetal or juvenile features that disappear in apes as they grow to adulthood, but persist in humans right up to death. The flatter faces and high foreheads that we and infant chimps share, for example. Our lack of body hair compared with chimpanzees and gorillas (fetal apes have little body hair). The form of our ears, the absence of large brow ridges over our eyes, a skull that sits facing forward on our necks, a straight rather than thumblike big toe, and the large size of our heads compared with the rest of our bodies. The list is long and Bolk's observations were absolutely accurate.* You can find every one of these traits in fetal, infant, or toddling apes, and all modern human adults. No less than evolutionary biologist Stephen Jay Gould agreed with Bolk

---

*See pages 37–38 of *Thumbs, Toes, and Tears: And Other Traits That Make Us Human* for Bolk's complete list.

in his own landmark book, *Ontogeny and Phylogeny* (though he didn't agree with the elder scientist's reasons for coming to those conclusions, which were tainted with racism and convoluted views of evolution). Gould called our peculiar brand of neoteny one of the most important twists in all the turns that human evolution has taken.[4]

Given its dictionary definition, you might think that neoteny is simply a matter of a species holding on to as many youthful traits of an ancestor as long into adulthood as possible (a little like Joan Rivers or Cher). But it's not that simple. Undeniably, in some ways we are childlike versions of our pongid ancestors, but in others our maturity is accelerated, rather than stunted. For example, while our faces and heads may not change as radically as an ape's as we enter adulthood, our bodies still continue to grow and change. We don't retain the three-foot stature of a two-year-old toddler. In fact at an average (worldwide) male height of five feet nine inches, give or take a few centimeters, we are among the largest gracile apes to have ever evolved. Nor is our sexual maturity slowed, though it is delayed compared with other human species (including Neanderthals, as we will see soon). And our brain development is anything but arrested. In fact, just the opposite. As I said, complicated.

The different ways some parts of us seem to accelerate and mature while others bide their time or halt altogether has generated a flock of terms related to neoteny—*paedomorphosis, heterochrony, progenesis, hypermorphosis,* and *recapitulation.* The debate is ongoing about what exactly *neoteny* and the rest of all of these labels truly mean. In the end, however, it comes down to this—each represents an evolution of evolution itself, an exceptional and rare combination of adaptations that changed our ancestors so fundamentally that it led to an ape (us) capable of changing the very planet that brought it into existence.[5] Put another way, it changed everything.

Mostly we think of Darwin's "descent by natural selection" as a chance transformation of newly arrived mutations—usually physical—into an asset rather than a liability, which is then passed along to the next generation. So paws become fins in mammals that have taken to the sea. The spindly arms of certain dinosaurs evolve into the wings of today's birds. The ballasting bladders of ancient fish become the predecessors of land animals' lungs. All of that is true. But what neoteny (and paedomorphosis and all the rest) illustrate is that the forces

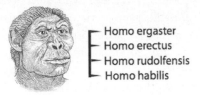

Homo ergaster
Homo erectus
Homo rudolfensis
Homo habilis

of evolution don't simply play with physical attributes, they play with time, too, or more accurately they can shift the times when genes are expressed and hormones flow, which not only alters looks but behavior, with fascinating results.

Evolution manages this by not affecting solely *what* traits it reveals, but *when* it reveals them. It moves abilities, physical features, and behaviors forward or backward, or stops them altogether by altering the expression of genes that affect developmental hormones. It plays with time like a boulevard-game master plays a shill game with walnut shells and peas. So in us, our big toe remains straight throughout our lives rather than crooking thumblike before birth as it does for chimps and gorillas. We remain relatively hairless, like fetal apes.[6] Our jaws stay square and our foreheads flat throughout our lives rather than sloping backward as we leave our early years behind. And instead of decelerating brain growth after birth like orangutans, chimps, and gorillas, the genes that control the amount and interconnections of neurons act as though we are still in the womb and continue to fervently multiply.

Put another way, after birth, processes that were once *pre*natal in our ancestors become *post*natal in us. By being born "early," our youth is amplified and elongated, and it continues to stretch out across our lives into the extended childhood that makes us so different from the other primates that preceded us. We see it in the fossil record. Almost without exception, the dusty bones scientists have unearthed and fitted together reveal that the faces of gracile primates such as *habilis*, *rudolfensis*, and *ergaster*, while still plenty simian, grew step by step to increasingly resemble us. Their snouts were flattening, their foreheads were growing higher and less sloped, their chins stronger. Features that once existed only in fetal forest apes like big toes and heads that rested upright on shoulders now not only existed in youth but also persisted into adulthood.

Exactly how all of this unfolded on the wild and sprawling plains of Africa isn't clear precisely, but there can be no doubt that it did. We stand as the indisputable proof. All of the evidence emphatically points to our direct, gracile ape ancestors steadily extending their youth. They were inventing childhood. But most important, to us at least, in the inventing they were becoming more adept at avoiding extinction's sharp and remorseless scythe. And the main reason that was happening was because the childhood that was evolving enabled the development of a remarkably flexible brain. That is where the grand story of our evolution made an extraordinary turn.

The clustered neurons that together compose the brains of all primates grow at a rate before birth that even the most objective laboratory researcher could only call exuberant, maybe even scary. Within a month of gestation primate brain cells are blooming by the thousands per *second*. But for most species that growth slows markedly after birth. The brain of a monkey fetus, for example, arrives on its birthday with 70 percent of its cerebral development already behind it, and the remaining 30 percent is finished off in the next six months. A chimpanzee completes all of its brain growth within twelve months of birth. You and I, however, came into the world with a brain that weighed a mere 23 percent of what it would become in adulthood. Over the first three years of your life it tripled in size, continued to grow for three more years until age six, underwent massive rewiring again in adolescence, and finally completed most, but not all, of its development by the time you reached your second decade (assuming that as you read this you *have* reached your second decade).

Being born so "young," you might conclude our brains arrive comparatively underdeveloped at birth, but that is not the case. Despite our early arrival we still come into the world bigheaded, even compared with our more mature cousin primates. At birth the brains of apes constitute 9 percent of their total body weight, hefty by the standards of most mammals. We, however, weigh in at a strapping 12 percent, which makes our brain 1.33 times larger than an infant ape's, relatively speaking, despite our abbreviated sojourn in the womb. In other words even arriving in our early, fetal state, with less than a quarter of our brain development under our belts, we are still born with remarkably large brains.

Keep in mind that this approach to brain development is so extraordinarily strange and rare that it is unique in nature. And dangerous. If an engineer were planning the optimum size of a brain at birth, it would clearly be illogical to bring newborns into the world this cerebrally incomplete. Too fragile, and too likely to fail. Far more practical to do all the work in the safety of a mother's body. But evolution doesn't plan. It simply modifies randomly and moves forward. And in this case, remember, remaining in the womb full term was out of the question. For us it was be born early, or don't be born.

As much as we might like to know the answer, exactly when it became necessary for our ancestors to exit the birth canal "younger" is frankly impossible to say. Since we *Homo sapiens* are the only human species to still be walking the planet since Africa's retreating jungles orphaned the rain-forest apes that preceded us, and since the skeletal remains of those who came before us are rare and difficult to decipher, we simply haven't yet gathered enough clues to know precisely when an early birth became unavoidable. There are, however, a few theories.

Some scientists believe earlier births would have begun when the adult brain of some predecessor or another reached 850cc.[7] Anthropologist Robert D. Martin calls this the "cerebral Rubicon," a line that once crossed would have required that some sort of longer, human-style childhood become part of that creature's life. If that's true, that narrows the candidates to those human species living between 1.8 and 2 million years ago—species like *Homo rudolfensis* or *Homo ergaster*. Until recently scientists felt *Homo habilis* (Handyman) was the best candidate, but new evidence has caused some realignment of the human family tree. For decades the common wisdom had it that we descended from *Homo habilis* by way of *Homo erectus*, which in turn evolved into what paleoanthropologists call "anatomically modern humans" (AMH), our kind. But new fossil finds now indicate that *erectus* and *habilis* were East African contemporaries for nearly a half million years, making it rather difficult to have descended from one another. Furthermore, *ergaster* and *rudolfensis*, which were often tossed in with *Homo erectus*, are now more often considered to be their own separate species.

This means that in the ever-shifting drama (and nomenclature) of human evolution, Handyman now represents an evolutionary dead end and *Homo erectus* may turn out to be not one species, but many, with only one particular representative leading directly to us, if that.

Whatever the case, around this time, when humans began to grow adult brains about three quarters of the size that ours are today, the offspring of upright walking humans may have been forced to arrive prematurely as the fit between head and pelvis grew increasingly tight. Who, the question then becomes, were the people from whom we directly descended, and where can we suppose they lived?

Some history might be in order.

Thirty-five million years ago the northeast corner of Africa was being carried on the back of a tectonic plate determined to make its way eastward toward Asia, while the rest of the continent was steadfastly refusing to go along. One consequence of this dogged parting of the ways was the emergence of the Arabian Sea and peninsula (with, coincidentally, all its oil beneath it). Another was the formation of a long and immense lake in East Africa made possible by the three substantial rivers that drained into it from the surrounding mountains. Two million years ago, the evidence of this great African rift, and the lake it created, was still all around. The ruptured land had left behind dozens of volcanoes, smoldering ominously and erupting unpredictably. One even rose defiantly out of the great lake itself, a disdainful sentinel that stood unfazed by the storms that howled when the seasons changed or the dust devils that spun along its flanks in the hot summer months.

If you check a map of Africa today, you will notice the slender imprint of this lake we now call Turkana (formerly known as Lake Rudolf). It is still vast, a long, liquid gem that lies on the breast of East Africa, most of it in northern Kenya with just its upper nose nudging the highlands of southern Ethiopia. Today Lake Turkana fails to be as hospitable as it was earlier in its life. The rivers that once drained it are gone, so evaporation is the only exit for Turkana's waters. That has turned it a splendid jade color and made it the world's largest alkaline lake. These days the land that surrounds it is mostly dry, harsh, and remote. However, 1.8 million years ago it was an exceedingly fine place to set up housekeeping.

Life of every kind thrived along Turkana's shores in the early days of the Pleistocene epoch, despite the occasional ferocity of the weather and the ominous belching of its volcanoes.[8] Crocodiles bathed in its warm waters; *Deinotherium*, an ancient version of the elephant, and both black and white rhinoceroses grazed among the grasslands.

Hyenas yelped and hooted, scavenging what they could and hunting flamingos that fed in the shallows, while the grandcousins of lions, tigers, and panthers harvested dinner from herds of an early, three-toed horse called *Hipparion*. The lake, the streams and the rivers that fed it, and the variability of the weather made the area a kind of smorgasbord of biomes—grasslands, desert, verdant shorelines, clusters of forest and thick scrub. The bones of the extinct beasts that lie by the millions in the layers of volcanic ash beyond the shores of Lake Turkana today attest to its ancient popularity.

Homo ergaster

The existence of a habitat this lush and hospitable wasn't lost on our ancestors any more than it was on the elephants, tigers, and antelope that roamed its valleys. In fact it was so well liked that *Homo ergaster* (left), *Homo habilis*, and *Homo rudolfensis* were all ranging among its eastern and northern shores 1.8 million years ago, sharing the benefits of the basin with their robust cousin *Paranthropus boisei*. As many as a million years earlier, *Paranthropus aethiopicus* came and went along the northwestern fringe of the lake, and half a million years before that the flat-faced one, *Kenyanthropus platyops*, braved Turkana's winds and watched its volcanoes rumble and spew.

Despite decades of sweltering work, paleoanthropologists have yet to categorically determine which of these humans who trod the shores of Turkana led directly to us, but it is possible to make an informed guess, at least based on the limited evidence scientists have to work with. We already know *Homo habilis* is out of the question, an evolutionary dead end unrelated to *Homo erectus*. *Homo rudolfensis* is also unlikely because he bears such a strong resemblance to *Paranthropus boisei* and his robust ancestors. He may have been a bridge species of some sort. *Boisei* himself would seem not to qualify given that he wasn't gracile (we are) and possessed the smallest brain of the group, the largest jaws, and the most apelike features.

That leaves *Homo ergaster*, "the worker" (*ergaster* derives from the Greek word ἐργαστήρ, meaning "workman"), formerly considered an example of *Homo erectus*. Truthfully, *ergaster* wouldn't seem to be a promising candidate for a direct ancestor either, except for one remarkable fossil find that has been, after some heated debate, assigned to the *ergaster* line. In the scientific literature he is known as Turkana (or sometimes Nariokotome) Boy because Kamoya Kimeu, a paleoanthropologist who was working at the time with Richard Leakey, came across him on the western shore of Lake Turkana.

His discovery first stunned his fellow anthropologists and then the world with the completeness of what he had found. In a scientific field where scraping up a tooth or a jaw fragment, or a wrecked piece of tibia, can be cause for wild jubilation, Kimeu and his colleagues uncovered not only a skull, but a rib cage, a complete backbone, pelvis, and legs, right down to the ankles. There, in the brittle detritus of the Dark Continent, lay the nearly complete remains of a boy who had lived 1.5 million years ago and died in the swamps of the lake somewhere between the ages of seven and fifteen. It was nothing short of remarkable.

You may have noticed the wide range of the boy's possible age. There's a reason for that. Despite being among the most studied fossils in the annals of paleoanthropology, scientists cannot seem to universally agree on the age of their owner, a mystery that brings us back to the issue of long and lengthening childhoods. The boy's age is elusive because we have only two living examples of primates that we can use as benchmarks to determine his age when he died—forest apes and us. But Turkana Boy is neither. With an adult brain that would likely have been about 880 cc, he falls almost midway between the two extremes. Take away half the mass of his brain, and it would be about the size of a chimpanzee's. Add the same amount and he would be within the range of most modern humans.

When scientists first inspected the boy's fossilized teeth, they immediately realized he was, in fact, a boy because several of them had not yet entirely arrived. In his lower jaw a few permanent incisors, canines, and molars had formed, but not all of them were fully grown. In his upper jaw he still had his baby, or milk, canines and no third molar. If a dentist were looking into a mouth like that today, she would conclude she was dealing with an eleven-year-old. But if the mouth belonged to a chimpanzee, seven would be a better guess.

Teeth represent one type of clue scientists use to help estimate the age of a skeleton (or more precisely, the skeleton's former owner) when he died. Another is growth plates. Long bones like those in our arms and legs don't fuse permanently with the joints attached to them until they are fully grown. The state of growth plates reliably predicts age. Turkana Boy's long leg bones were still growing and had not yet fused, particularly at the hips, although one of his shoulder and elbow joints was fusing. Given the state of his growth plates, researchers concluded the boy could have been as young as eleven or as old as fifteen the day he met his untimely end, *if* he was human. Or a mere seven if he was a chimpanzee.

A final feature that helps determine age is height. Nariokotome Boy's thighbone is seventeen inches long, which would have made him roughly five feet three inches tall, about the size of an average fifteen-year-old *Homo sapiens*, or a full-grown chimpanzee. Compared with other fossil primates, australopithecines or even his Turkana contemporaries like *Homo habilis* and *rudolfensis*, for example, Nariokotome was tall, and depending on his exact age, he might have grown considerably taller, had he survived. So how old *was* the "working" boy?

Viewed from either end of the spectrum, none of the clues about his age have made much sense to the teams of scientists who have labored over them. Each was out of sync with the other. Some life events were happening too soon, some too late, none strictly adhering to the growth schedules of either modern humans or forest apes. Still, the skeleton's desynchronized features strongly suggested that the relatives of this denizen of Lake Turkana were almost certainly being born "younger," elongating their childhoods and postponing their adolescence. Apes may be adolescents at age seven and humans at age eleven, but this creature fell somewhere in between.

If the Rubicon theory is correct, and an adult brain of 850 cc marked the time when newborns begin to struggle to successfully make their journey through the birth canal, *ergaster* children were likely already coming into the world earlier than the rain-forest primates that preceded them five million years earlier. On the other hand, Turkana Boy was not being born as "young" as we are. His large brain, as large as any other in the human world at that time, and his slim hips, optimized for upright walking and running, reinforce the evidence. He must have been born "premature" or he wouldn't have been born at all. But if he was being born earlier, how much earlier?

Suppose the brain of a fully grown Turkana Boy was 60 percent the size of our brain today. (We have to suppose because we have no adult *ergaster* cranium to consult.) And let's assume *ergaster* children would have come into the world after fourteen months of gestation, approximately 30 percent sooner than a chimp. This isn't as drastically different as the eleven-month disparity between other primates and us modern humans, but it would have represented the beginning of a significant human childhood, and it would have begun to upend the daily lives of our ancestors in almost every way.

Why? First, there would have been more death in a world where, unfortunately, death was no stranger. Many "early borns" would have died after birth, unable, unlike today's chimps and gorillas, to quickly fend for themselves. Because gorilla and chimp newborns are more physically mature than human newborns, they often help pull themselves out of the birth canal and quickly crawl into their mother's arms or up onto her back. It's unlikely that *ergaster* infants were capable of this. Of all primates, human newborns are by far the most helpless. When we arrive, we are utterly incapable of walking or crawling. We can't see well or even hold our heads up. Without immediate and almost constant care, we would certainly die within a day or two. Though these "preemies" were not likely as defenseless at birth as we are, they would have been far less physically developed than their jungle or even early savanna predecessors.

But even if the newborns didn't die in childbirth, their mothers might have, their narrow pelvises unable to handle what scientists call the expanding "encephalization quotients" of their babies. To compensate, *ergaster* newborns may have begun to turn in the birth canal so that they were born facedown, a revolutionary event in human birth. Unlike other primates, our upright posture makes it necessary for babies to rotate like a screw so they emerge facedown. If they came out faceup as chimps and gorilla infants do, their backs would snap during birth. Altogether, this made an already difficult process even more dangerous and difficult.

The job of bringing a child into the world would not only have become more complicated, but imagine life for the mothers of these offspring, assuming both survived the ordeal of birth. They were already living a precarious existence in a menacing world—open grasslands or at best thick brush with occasional clusters of forest. Predators such as striped hyenas and the sythe-toothed *Homotherium* had appetites and needs, too. There was no such thing as a campfire to keep predators at bay. Fire had yet to be mastered. When night fell, it was black and total with nothing more than the puny illumination provided by the long spine of the Milky Way, a fickle moon, or an occasional wildfire in the distance sparked suddenly and inexplicably by lightning or an ill-tempered volcano. And the big cats of the savanna like to hunt when the sun has set.

Not only were these new human infants more helpless than ever, but their neurons were proliferating outside the womb at the same white-hot rate they once did inside. Rapidly growing brains demand serious nutrition. Studies show that children five and younger use 40 to 85 percent of their standing metabolic rate to maintain their brains. Adults, by comparison, use 16 to 25 percent.[9] Even for *ergaster* children, a lack of food in the first few years of life would often have led to premature death. Nariokotome Boy might have been undernourished himself. His ancient teeth reveal he was suffering from an abscess. His immune system may not have been strong enough to defeat the infection, and lacking antibiotics, scientists theorize blood poisoning abbreviated his life. He was probably not the first among his kind to die this way.

In every way, early borns would have made life on the savanna more difficult, more dangerous, and more unpredictable for their parents and other members of the troop. So why should evolution opt for larger brains and earlier births? And how did it manage to make a success of it?

Difficult question to answer. Looking back on the scarce orts of information science has so far gathered together, premature birth doesn't make an ounce of evolutionary sense. Not on the surface. Darwinian adaptations succeed for one reason—they help ensure the continuation of the species. That means if your kind misplaces the habit of living long enough to have sex successfully, extinction will swiftly follow. Since this is the ultimate fate of 99.9 percent of all life on earth, it is difficult to fathom how the mountains of challenge that early-arriving newborns heaped on the backs of their gracile ape parents could possibly help them successfully struggle to stay even a single step ahead of the grim reaper.

It certainly wouldn't seem to make much sense to lengthen the time between birth and sex. Keeping that time as brief as possible has immense advantages after all. It's a powerful way to maximize the number of newborns either by having large numbers of them at once or by having them often, or both. Dogs, for example often enter the world in bundles of five or six at a time, are weaned by six weeks, and ready to mate as early as six months. They aren't puppies long, and once they are done breast-feeding, they are soon prepared to fend for themselves. For mice the process is even more compressed. The result is that mothers bear more children with every birth, do it more often, and those offspring are quickly ready to mate and repeat the cycle. All of this accelerates the proliferation of the species *and* improves its chances of survival.

We humans, however, wait an average of nineteen years before bearing our first child. Why? If shortening the time between being born and bearing as many offspring as often as possible works so well for other mammals, for what reasons would evolution twist itself backward with Africa's struggling troops of savanna apes? Why bring increasingly defenseless infants into the world? Why expose their parents to greater danger to feed and protect them? Why insert this extra, unprecedented cycle of growth, this thing we call childhood, into a life—a time when we rely utterly on other adults to take care of us? And what advantage is there in taking nearly two decades to bring the first of the next generation into the fold?[10]

In his landmark book *Ontogeny and Phylogeny*, Stephen Jay Gould spends considerable time discussing two types of environments that drive different varieties of evolutionary selection. One he calls *r* selection, which takes place in environments that provide plenty of space and food and little competition. A kind of animal Valhalla. The other is called *K* selection, environments where space and resources are scarce and competition is nasty and formidable.

R selection (Gould points out many studies that back this up) encourages species to have plenty of offspring as quickly as possible (think rabbits, ants, or bacteria) to take advantage of the lavish resources at hand. But *K*-style environments require species to slow down, create fewer offspring, and take more time doing it because it reduces stress on the environment and the competition among those trying to survive in it. By random chance, evolution begins to favor the creation of fewer competitors within a species who will only die off from lack of resources.[11] By reducing death and lengthening life, in particular early life, *K* selection also provides species extra time to develop in ways that make them more adaptable. In our case, as Gould put it, *K* selection made us "an order of mammals distinguished by their propensity for repeated single births, intense parental care, long life spans, late maturation, and a high degree of socialization." Today you and I stand as the poster children for K-strategy evolution. Yet, while the simple fact that we are walking around today provides conspicuous proof that K strategies can succeed, it still fails to explain *why* they succeed.[12]

It is possible that it didn't, at least not all the time. Multiple species arguably walked down this Darwinian road and were snuffed out. Several—about whom we may never know a thing—were surely done in over time by the unrelenting pressures of protecting their helpless infants, braving their environment to get them more food, or becoming dinner themselves for some salivating savanna cat. Is this what wiped out *Australopithecus garhi*? Does this explain the demise of *Homo habilis* or *rudolfensis*? So far the sparse, silent, and petrified clues that the fossil record has left us aren't parting with those secrets. They are stingy that way.

We do know this: around a million years ago or so—early November in the Human Evolutionary Calendar—the robust primates had met their end, and so had many gracile species, but a handful continued and even flourished. Already some had departed Africa and had begun fanning out east to Asia and the far Pacific. The cerebral Rubicon had been crossed and there was no going back.

This meant that evolution's forces had opted, in the case of our direct ancestors, for bigger and better brains rather than more sex and more offspring as a survival strategy. And, against all odds, it was working—a profound evolutionary shift. Over time, in the crucible of the hot African savanna, far away in time from the Eden of rain forests, an exchange was made—reproductive agility for mental agility. If bringing a child into the world "younger" was what it took, fine. If expending more time and energy on being a parent was necessary to ensure that a creature with a bigger, sharper brain would survive, then so be it. If evolving an entirely new phase of life that created the planet's first children was required, then it had to happen. The imponderable forces of evolution had made a bet that delivered not greater speed or ferocity, not greater endurance or strength, but greater intelligence, or put in flat Darwinian terms, greater adaptability. Because that is what larger, more complex brains deliver—a cerebral suppleness that makes it possible to adjust to circumstances on the fly, a reliance not so much on genes as on cleverness.

It is strange to think that events could well have gone another way. Earth might today be a planet of seven continents and seven seas and not a single city. A place where bison and elephants and tigers roam unheeded and unharmed, and troops of bright, robust primates live throughout Africa, maybe even as far away as Europe and Asia, with not a single car or skyscraper or spaceship to be found. Not even fire or clothing. Who can say? But as it happened, childhood evolved, and despite some very long odds, our species found its way into existence.

CHAPTER THREE

LEARNING MACHINES

It is easier to build strong children than to repair broken men.
—Frederick Douglass

Boy, n.: a noise with dirt on it.
—Anonymous

Give me the child until he is seven, and I will give you the man.
—Jesuit aphorism

YOUTH, THE FRENCH writer François Duc de La Rochefoucauld once observed, "is a perpetual intoxication; it is a fever of the mind." Ralph Waldo Emerson was more blunt: "A child is a curly, dimpled lunatic." We have all witnessed a toddler or two in action (usually and most memorably our own), and it is a sight to see. The average two-year-old is thirty inches tall and twenty-eight pounds of pure, cerebral appetite determined without plan or guile to snatch up absolutely everything knowable from the world. She, or he, is, indisputably, the most ravenous, and most successful, learning machine yet devised in the universe.

It may not seem so on the surface, but toddlers accomplish prodigious amounts of work (all cleverly disguised as play) as they bowl and bawl their way through each day. By throwing a ball (or food), playing in the mud, a pool, or a sandbox, by attempting headlong runs and taking sudden tumbles, or swinging on swings or off "monkey" bars, the young are fervently familiarizing themselves with Newton's laws of motion, Galileo's insights into gravity, and Archimedes' buoyancy principle, all without the burden of a single formula or mathematical term.

When a toddler smiles, cries, grimaces, gurgles, giggles, spits, bites, or hits; when she breaks free of mom or dad for a wild dash down the sidewalk; throws whatever he can grab for the sheer joy of it; dances spontaneous jigs or engages in other diabolical antics—she is learning what is socially acceptable and what is not, what is scary, what works in the way of communication and what fails and when. Food, and much that isn't food, is tasted, licked, and baptized with slobber to investigate its texture, shape, and taste. Yet no artifice or logic is behind the tasting. It's just another form of exploration. Objects, living or not, are bounced, swatted, hugged, flailed, closely inspected, all in a fervent effort to comprehend their nature. Unbounded and unstoppable greed for knowledge is the best way to put it.

Acquiring language is another big job in childhood. Babbling, squeals, and other noises are, as the best linguists have so far been able to ascertain, ways of figuring out the language that the other bigger, parental creatures speak around the toddlers whether it is Swahili, German, or Hindi. Later, early conversations are short, generally. "Here! Mama! Dadda! Mine! No! Please. Want!" Often, in times of acute frustration, communication is inarticulate, loud, and punctuated with acrobatic body language. In time, however, and amazingly, vocabularies grow, syntax improves, and full sentences are expressed, all with hardly an ounce of formal instruction. The acquisition of language is one of the great miracles in nature. At age one few children can say even a single word. At eighteen months they begin to learn one new word roughly every two hours they are awake. By age four they can hold remarkably insightful conversations, and by adolescence they have gathered tens of thousands of words into their vocabulary at a rate of ten to fifteen a day and often use them with lethal effect! And nearly every word was acquired simply by their listening to, and talking with, the people around them.[1]

Children do these apparently lunatic and astonishing things for a reason. Nature has wired their brains for survival by driving them to swallow the world up as fast as they possibly can. Pulling off this feat is easier said than done. However, if we hope to comprehend how cerebral connections this complex take place, it might first be useful to step back and consider why brains exist at all, and how we eventually came by the particular brand we have.

★　★　★

By general agreement the first brain in nature belonged to a creature scientists today call planaria, known more commonly to you and me as the lowly flatworm. Flatworms are metazoans and wouldn't seem therefore to be very brainy. But intelligence is a relative thing and planaria, when they first emerged more than seven hundred million years ago, were the geniuses of their time, creatures of unparalleled intelligence blessed with an entirely new kind of sensory cell capable of extracting marvelously valuable bits of information from their environment.

Unlike many of their contemporaries planaria were unusually sensitive to light, possessed rudimentary sets of eyes, and responded to, rather than ignored, changes in temperature—all radical innovations in their time. Even today they remain expert at sensing food, and then making their way with uncanny determination to it, while other metazoans (corals, for example) generally take a more leisurely approach to their cuisine, waiting for it to find them rather than the other way around.

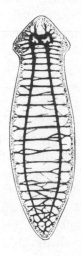

Planaria—the Einstein of the Day

Among the cellular innovations that made an ancient flatworm's brain possible was a protoneuron called a ganglion cell. These are clustered in the head of the worm and then connected to twin nerve cords that run in parallel down the length of its body so that certain experiences sensed alongside it can be transmitted to the flatworm's brain for some metazoan cogitation. All the brains that evolution has so far contrived rest on this tiny foundation. So for the best ideas you had today you can thank the determined metazoan that looks something like a squished noodle.[2]

The purpose of brains generally is to organize the waves of sensory phenomena that nature's cerebrally gifted creatures experience. Their job is to filter the world's chaos effectively enough to avoid, for as long as possible, the disagreeable experience of death. A direct correlation exists between survival and how well a brain maps the world around it. The more accurately it can correlate, the more likely it will survive danger, discover rewards, and keep its owner among the living.

At the heart of every brain are its neurons, the specialized cells that make possible our brand of thinking, feeling, seeing, moving, and nearly everything else important to us. There are over 150 different kinds of neurons, making them the most diverse cell type in the human body. To support their greedy habit of consuming large quantities of energy, they are surrounded by clusters of glial cells, which serve as doting nannies busily shuttling nutrients and oxygen to them while fetching away debris and generally working to keep the neurons fresh and firing. Each of us carries roughly a hundred billion neurons clustered jellylike inside our skulls (coincidentally about the same number as stars cosmologists believe populate the Milky Way galaxy). Every one of them is supported by ten to fifty indulgent glial cells.

This makes our brains a remarkable and mysterious place still well beyond the comprehension of the thing itself (a fascinating irony), but the cerebral cortex of a growing human child is more remarkable still. Only four weeks after a human sperm and egg successfully find one another, when we are still embryos no larger than a quarter, clusters of neurons that will eventually become our brain are replicating at the rate of 250,000 every minute, furious by any standard. Around this time, a bumpy neural tube that looks suspiciously similar to a glowworm has begun to take shape. Over the next several weeks four buds within the tube will begin developing into key areas of the brain: the olfactory forebrain and limbic system—the seat of many of our primal emotions; the visual and auditory midbrain, which governs sight, hearing and speech; the brain stem, which controls autonomic bodily functions such as breathing and heartbeat; and the spinal cord, the trunk line for brain–body communication. Two weeks later a fifth cluster of neurons begins to blossom into the frontal, parietal, occipital, and temporal lobes of the cerebral cortex, where so many exclusively human brain functions reside.[3]

The brain constructs itself this way, with neurons ebulliently proliferating, and then, like the rest of the cells in the embryonic body, they
march off to undertake their genetically preordained duties. During
this process and throughout our lives, every cell in the body commu-
nicates. It is in all cells' DNA, not to mention our best interests, to
reach out and touch one another, mostly by exchanging proteins and
hormones. But neurons are especially talented communicators. This
is because whenever the biological dice fell in such a way that they
came into existence, they began to evolve specialized connectors—
dendrites and axons—that vastly improved their exchange of information compared to other cells in the body.

Before brains came along, primitive protoneurons communicated
by secreting hormones and electrical currents in no particular direction, mumbling their messages to the other cells and protoneurons in
their vicinity, and not getting terribly quick results, at least compared
with our current models. With the invention of dendrites and axons,
however, they could form elegant, smart clusters that shared at high
speed the information each of them held with the others nearby. (Planaria were among the first to accomplish this.)

The emergence of high-speed, if exceedingly minute, communications cables meant that any creature fortunate enough to inherit them
could more fully and rapidly sense the world it inhabited—light and
dark, food, danger, pain and pleasure—then react to it all in a blink.
Not only that, the cables could link different sectors of the brain the
way highways connect cities. This meant the brain could not only improve contact with the world, but also stay in better touch with itself,
not a trivial matter as brains grew larger. (This turns out to be important to consciousness, but we will visit that subject later.)

Dendrites generally conduct signals coming into a brain cell while
axons do the opposite. Dendrites (also known as dendrons) are so eager to make contact that they extend treelike in multiple directions
and can place one neuron in touch with thousands of its neighbors.
Axons aren't nearly as obliging as dendrites, but can still make uncounted connections as they transmit signals outward when a neuron
is stimulated and reaches what is known as a threshold point, a moment that is vitally important when it comes to thinking, feeling, and
sensing. At that instant an electrical impulse bolts down the axon at
270 miles an hour. When it reaches the end of the axon, a tiny pouch
of chemicals bursts, sending neurotransmitters across a synaptic gap

like party confetti, where they embrace the receptor sites of the next
neuron like a long-lost relative and then pass along their message.

Your brain is capable of making one quadrillion (that's a 10 with
fifteen zeros behind it) connections like these. Even as you read the
words in front of you, impulses are flaring out and back at high speed,
a three-dimensional, electrochemical storm tirelessly at work conjuring
your thoughts, assessing your feelings, ensuring your body operates according to plan, and generating your personal version of reality. It's a
busy place.

While neurons multiply at blistering rates before we are born, the business of building the brain continues even more earnestly after we enter
the world. By strict decree, the twenty-five thousand genes—the
"structural genome"—each of us inherits in fifty-fifty doses from our
parents resolutely continue the construction of our own wetware, and
its underlying neuronal infrastructure, complete with our specific talents and predispositions. Just as some of us may inherit stocky bodies
and others long, slim ones, our parents can also issue brains that incline
us to be gregarious or shy, a leader more than a follower, mathemati-
cally, musically, or verbally predisposed. This part of us is a genetic
crapshoot, and we have no control over it.

Nevertheless, more than other forms of life, even other primates,
we can be thankful that we are not immutably linked to our genetic
directives. In us they are editable, able to be altered by our personal
experience and environment, a phenomenon that explains why each
of us is not a clone of the other, not even in the case of identical twins,
who carry precise copies of their sibling's DNA. It is impossible to
overemphasize the impact this new ability had on human evolution
and has each day on your life and mine. The farther down the evolutionary chain creatures fall, the less complex their brains are as a rule,
and the less they are shaped by their personal experience, which is
another way of saying that their day-to-day actions are largely, if not
entirely, governed by their genes, rather than by anything we might
call a "self."

Moths, for example, are drawn to candle flames because they are
genetically programmed to navigate by the light of the moon. Not
having much of a brain, they have been known to mistake a flame for
the moon and get incinerated for their trouble. This happens not sim-

ply because their brain is small, but because it is also hardwired by its genes and not readily able to learn from experience.

For hundreds of millions of years genes were a perfectly effective, if plodding and random, way of adapting to changes in environment, but it wasn't efficient. It took a long time for evolution to get around to building a brain that could think, even a little, for itself. But once it did, those animals blessed with one tended to survive longer than those that weren't. Brains are more resourceful than trial-by-error genetics. They map the world in real time and increase the chances that you will make a lifesaving decision on the spot rather than a deadly, DNA-dictated one that isn't even aware you *are* on the spot. Not that the influence of genes versus brains is either/or. All creatures endowed with a brain lie along a continuum of cerebral, and therefore behavioral, flexibility. There are no hard boundaries. But the *degree* of that hardwiring in many ways marks the difference between, say, a flatworm, and us.

The impact that the outside world can have on our brains during our childhood explains how seven billion of us can be walking the planet every day, each a thoroughly unique universe unto ourselves, distinct in personality, experience, thought, and emotion; yet similar enough that we can (more or less) relate to one another and be counted as members of the same species. What has been far less clear, and a slippery problem for scientists, has been exactly how the genetic commands we inherit from our parents are bent by the unique relationships and events in our lives. It turns out several forces are at work. Very hard at work.[4]

In the first three years of life the human cerebral cortex triples in size. This is like nothing else in nature. Yet it isn't simply the growth of neurons that makes the human brain so powerful. It is also the way it feverishly links them up. Why should this matter? Think of the brain as a miniature, though considerably more complex, Internet, compressed in size and time. Each neuron is like a computer sitting on a lap or desk somewhere. Computers today are powerful, like neurons, and can by themselves accomplish a great deal. I am writing this book on one right now. But connect neurons or computers to one another, and they become amplified and add up to far more than the sum of their parts. When my computer links to the Internet, it enables me to research information I use in the book, share passages I am writing with others in a blink, and gather opinions, thoughts, and insights by engaging in any number of conversations. I can instantly track down specific bits of information I need or download facts, maps, images, even whole books and movies. By branching out and communicating in all directions, my computer becomes, in many ways, all the computers it can touch. Now multiply this by millions of sites from Facebook to the Library of Congress, billions of Web pages, and innumerable other computers, and you begin to get a feel for the benefits of interconnecting neurons in the brain. There is power in communication.

The pathways between neurons begin to radiate almost the moment nerve cells undertake their growth in the fetal brain. Yet while the proliferation of neurons begins to slow at age three, the branching of pathways between them continues more urgently than ever. So urgently that a thirty-six-month-old child's brain is twice as active as a normal adult's, with trillions of dendrites and axons making contact, jabbering and listening and tightening the collaborative party that makes the human mind possible. One neuron can be directly linked to as many as fifteen thousand other nerve cells, generating more connections within the brain than there are electrons and protons in every heavenly body within every one of the hundred billion galaxies in the universe. That's a lot of communication, and it is all happening between your ears.

The culprits behind this mad construction project, the forces that create and shape these connections, are the boisterous circles of the outside world with all of its smells and sensations, sound, touches, social interactions, and dangers. In attempting to make sense of the world it lives in, the brain creates its connective architecture by smelting and hammering out a massive, riotous explosion of wetware, which is shaped by a child's sensory conversation with the world. The trillions of connections that blossom physically and chemically represent every new, frightening, exhilarating, or surprising experience children come across, which in the case of children is almost everything. For a toddler, novelty is riot in life. Since even big brains can't predict the future, this is nature's way of attempting to prepare for all flavors of trouble (and pleasure) yet to come; an all-out effort to create synaptic antennae that can better sense what may be, or could be, and use whatever tools and information are at hand to the best possible advantage. If music is part of your life, then neuronal pathways and structures begin to fan out to better handle, at first, listening to music, and then later making it. The same holds true for language, physical dexterity, sight, and social cues. Everything from the mundane to the sublime is shaped in the brain by the events around us.

You will have realized by now that this pretty much renders the old nature-versus-nurture debate irrelevant. The trillions of connections our brains make in childhood help explain why we are neither purely a product of our genes nor altogether the result of our personal experience, but both. Nevertheless, this does not represent the whole picture. The brain is like an onion. Peel back one mysterious layer and it only reveals another: a recently discovered parallel genetic system, for example, that works within each of us, and profoundly affects the person we become. This system is related to the genome, but it is not the genome. It's something else equally as fascinating called the epigenome.[5]

The long and spiraled strands of DNA that vibrate within the cells of all living things dictate whether they are plant or animal, have feet or wings, lungs or gills, and explain why you and I are tall or short, blond or brunette, Asian or black, even human as opposed to a planaria. But as if that weren't impressive enough, there is still more to our DNA. It is wrapped around proteins called histones. This two-leveled structure—the histones and the DNA—constitutes the epigenome. Scientists are a long way from fathoming the many-layered mysteries of epigenetics, but they know that when such a structure is tightly coiled around inactive genes, it renders them utterly silent and unreadable, but when it relaxes pressure the genes become more accessible and therefore more expressible. How exactly these genes are expressed depends on our personal experiences and the environment in which we live, physically, socially, and emotionally. Specific experiences can deeply affect different brain circuits during developmental stages that go by the self-descriptive term *sensitive periods*. Cells in different parts of the brain that affect sight, language, hearing, are sensitive at different times and for differing lengths of time in life, particularly childhood. How deeply our epigenetics change shapes the circuitry in our brains, which in turn shapes how we behave and who we are. Once a sensitive period passes, particular circuits grow set in their ways and then lie beyond the reach of new experience.

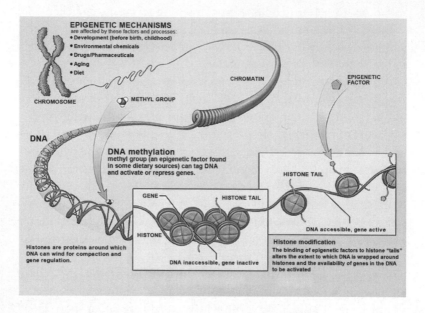

How the Epigenome Changes Your Brain

So while the codes in our DNA are set for life depending on what our parents pass along to us, we still have plenty of room to deviate from the precise commands of those genes. Thanks to the epigenome, events, and the physical and psychological environments in which we live during our childhoods, can modify the expression of some genes that affect brain development. Some of these amendments can be temporary; others can change us for the rest of our lives.

Study after study, for example, has found that children exposed to high stress are more likely to suffer mental illnesses later in life, including generalized anxiety and serious depression. High childhood stress has also been shown to modify how a person later handles adversity in adolescence and adulthood. When we are frightened, our adrenal glands release adrenaline, which focuses our attention, increases our heart rate, and prepares our bodies to either fight or flee, handy reactions when your life is on the line. But chronic fear and stress—the kind that continues relentlessly—can corrode us because intense, ongoing awareness of the flight-or-fight kind wears us out. In children an epigenome exposed to constant stress tends to make those children more sensitive to even minimal stress throughout their lives, and more likely to feel anxious when others might not feel the least bit nervous. Poor nutrition or toxic substances can affect epigenomes related to brain development during childhood in ways that blunt brain function later. Together these forces can gang up to have a kind of psychological domino effect that spills into our physical health to make us more susceptible to ailments like asthma, hypertension, heart disease, and diabetes.

On the other hand, positive experiences—warmth, stability, security, love, and the joy that comes from play—can create equally powerful, but entirely positive, results. Your genes write the basic blueprint of what is personally possible, or impossible. They set the boundaries of who you are physically, psychologically, socially, and intellectually, but your epigenome etches the finer details of your personality—the ways you handle others, your fears, joys your intellectual and emotional prowess, personal talents, confidence, proclivities for optimism or pessimism, and your annoying (not to mention altogether charming) quirks. They influence whether, when, and how your personal set of genes build the capacity for thought, emotional control, and a whole bushel of other future skills. Exactly what route the timing and depth of their effect takes depends on the infinitely complex molecular interactions that constitute your world and your "self." No matter what, the result is that you come out of it all as unique as a snowflake.

In case the connection has eluded you, it's our neotenous nature, our long childhoods, that makes our epigenome so inclined to the influences of our personal experience during the first seven years of life. Because we are born early and since we have extended our brain development well beyond the womb, neuronal networks that in other animals would never have been susceptible to change remain open and flexible, like the branches of a sapling. Although other primates enjoy these "sensitive periods," too, they pass rapidly, and their circuits become "hardwired" by age one, leaving them far less touched by the experiences of their youth. This epigenetic difference helps explain how chimpanzees, remarkable as they are, can have 99 percent of our DNA, but nothing like the same level of intellect, creativity, or complexity.[6]

As productive and interesting as all the goofy openness and flexibility of our toddlerhood is, it also creates a problem. It's not sustainable. Unless we hope to be a race of primates suffering from terminal cases of attention deficit disorder, the time eventually comes for our lively cerebral growth to be curbed. It's the biological equivalent of fishing or cutting bait. We can't afford to record, some way or another, every experience throughout our lives. The costs are too high. Our minds would grow so flexible they would become floppy, and so cluttered they would be incapable of focus. Besides, not every new experience is useful (sitting in traffic, for example). There are also physical limits to how big a brain can grow, though we *Homo sapiens* have certainly pushed the boundaries. Finally, brains, being greedy organs, devour immense quantities of energy for their size, especially in childhood. A growing toddler's cerebral appetite gobbles up as much as 85 percent of all the energy that its body requires each day. Over a lifetime that would be insupportable.

No, at some point the brain must make some tough biological choices by locking into and holding on to the influences the epigenome has expressed, while it somehow brings the sweeping array of connections it generates during childhood under control.

In the case of the epigenome, this process is relatively simple, if anything that happens in the human brain can be called simple. Despite the wild partying your youthful cerebral cortex undertakes, your genes are still ultimately in charge and dictate when different areas must calm down and mature. Genes decide when the sensitive periods of different cerebral circuits end, and when they end, that's that. It's interesting that it happens this way, almost as if each sector were a different brain, each with different genetic rules, which just happen to be expressed within the same skull (which in many ways is precisely the case, since different parts of the brain evolved at different times).

Neural circuits that analyze color, shape, and motion, for example, mature in the visual cortex long before higher-level functions develop to comprehend facial expressions, or the shape and meaning of frequently used objects, a glass, a fork, or a toy, for example.

For its part the auditory cortex first learns to recognize simple sounds, then later comprehends the meaning of those noises as words in strings of language that in turn help us make decisions or digest a great novel. The same process is true of other areas that handle physical and cognitive capabilities. Once these parts of the brain mature, the chances of changing them drop precipitously. Not that this is the absolute end of the epigenome's work. Throughout life, some matured areas will keep interconnecting to record revisions, mistakes, and the additions of knowledge that make us smarter, even wiser. But generally, it is during your childhood that the brain takes what the world has to offer and makes a bet that what its circuits have recorded is representative enough to handle what life will bring in the future. Following our childhoods, memory becomes our most effective way for us to change our behavior.[7] (In a way, memory is evolution's way of allowing you and me to remain in a permanently "sensitive period," always open to change, regardless of age.)

Controlling the other way that personal experience changes us—that would be the rampant connections our young brains make—is also related to the epigenome, but different from closing down sensitive periods. Remember, many of the pathways are created based on what we are exposed to as we grow, from music and language to sports and social interactions. You might assume that together they confer immense evolutionary benefits. They do, up to a point. But again there are limits to what we can handle. Too many connections make for a cluttered cerebellum.

The solution to this overabundance is a kind of intracranial evolutionary competition. Just as the organisms in an ecosystem are "selected out" if they can't find a niche where they can make their living, connections in our brains that are not used much after they are formed die out as well, unmasked as extravagances that have no place in the neurological ecosystems of our personal experience.

You can look at the first, rapid interconnections the brain makes based on experience as something like a matrix of dirt paths branching from your neurons, tentative explorations of this or that destination. The more often you undergo an experience—listen to music, catch a ball, hear a language, or are subjected to scary or stressful situations—the more often the path is walked and the more grooved it becomes. Paths may even become, metaphorically, paved or built into interstate highways, autobahns of thought and experience, because they are traveled so often. The highways remain for life, but in time, if the dirt paths and the back roads aren't traveled much, they disappear from lack of use. Only the well-traveled roads survive.

The synaptic routes we build, or not, deeply affect our perception and sense of reality. One of the most dramatic examples of this is the experience in the 1950s of anthropologist Colin Turnbull, who was researching the BaMbuti Pygmies, who live in the dense Ituri forests of central Africa. During his research he and a BaMbuti tribesman that he had come to know named Kenge traveled to another part of Africa characterized by broad plains, as opposed to the dense jungle Kenge had grown up in.

One day the two men were standing overlooking the expansive grasslands. Kenge pointed his finger at a herd of water buffalo and asked Turnbull, "What insects are those?" At first Turnbull couldn't figure what Kenge meant, then realized that he was referring to the buffalo. To this man, who had never before seen so much distance between anything, the buffalo appeared not small because they were far away, but small because, like an insect, they were, well, simply small. Turnbull realized that because the BaMbuti grow up in dense forests they never develop the ability to "see" or comprehend distance. When Turnbull told Kenge that the insects were buffalo, Kenge roared with laughter and told Turnbull not to tell such stupid lies. Given the world he grew up in, seeing things that were far away was a visual extravagance for Kenge, and so those connections to the visual cortex were paired away or, perhaps, never made at all.

If you were blindfolded between the ages of three and five, the same biology would be at work for you except in this case you would grow up entirely and forever blind. Not because your eyes are incapable of sight, but because the synaptic connections your brain made for sight before you were blindfolded would have been pruned away from lack of use by age five, the time when the visual cortex "hardwires" itself. The pathways would never have become paved highways because they were never used. Once that part of the brain locks in, there is no known way to restore sight. The pathways are gone. Children who suffer from amblyopia (lazy eye) often end up blind in their weak eye for the same reason. If the eye is rarely used, its connections to the visual cortex atrophy and die, even though the eye itself is perfectly functional.[8]

The most universal example of how our brains discard unused synaptic connections is language. Within five months or so of birth we all become capable of babbling every one of the sounds required to speak any of the sixty-three hundred languages humans utter throughout the world, and probably many that have long been extinct. Up to about age seven, the last year of childhood, children rapidly learn to speak in whatever tongue they are exposed to. If they encounter several, they will adopt them all with ease because to their brain the separate languages aren't separate at all, they are one; it's just that they have more words and rules.

Acquiring fluency in different languages later in life becomes more difficult because the neural circuits that help us master the sounds, the accents, and the grammar of those languages were never formed or have largely evaporated from lack of use. Even if you do learn a new language and get the grammar and the vocabulary right, it is nearly impossible to drop the accents you bring with you from your mother tongue. Henry Kissinger, for example, who has been speaking English since his teens, still speaks it with a strong German accent because English was not his first language. He didn't learn it until age fifteen, when his family migrated to the United States.

If the cerebral Rubicon theory is accurate, humans like *Homo ergaster* began being born "early" one million years ago. Their brains were now roughly three quarters the size of ours, so they weren't arriving in the world at a fetal stage as delicate as yours or mine, but the increased size of their heads was pushing them out of the womb sooner, and extending their childhoods. This meant they exited the womb uncompleted, a work in progress, hormonally primed to grow vast new farms of neurons and synapses, an amalgamation of their parents' genetic donations, but editable, more than any other creature up to that time, by their personal experience and the forces of the environment they faced.

There is no way to overestimate how important this was to our evolution. This was the birth of human childhood itself, and the beginning of wild and complicated processes that explain how you or I can be born in Fargo, North Dakota, learn to speak fluent French in Paris, develop wit like Woody Allen's, or become as reclusive as Howard Hughes, all while still plumbing the intricacies of subjects as wildly different as calculus, Mozart, and baseball. This began the trend that has, in many ways, made children of all of us the entire course of our lives, neurologically nimble enough that we can keep learning, changing, and overriding the primal commands of our DNA. As we age the pliability of our younger brains may grow more brittle, but they also become more stable, deeper, and broader. Or as anthropologist Ashley Montagu put it, "Our uniqueness lies in always remaining in a state of development."

If we stand back and gaze thoughtfully at the whole vista of human evolution, it grows clearer that the longer a childhood lasts, the more individualized the creatures that experience it become. It is the foundation of the thing we call our personalities, the unique attributes that make you, you and me, me. Without it, we would be far more similar to one another, and far less quirky and creative and charming. Our childhoods bestow upon us the great variety of interests and personalities and talents that the seven billion of us display all around the world every day from Barack Obama to Lady Gaga to Itzhak Perlman. This diversity led, slowly, to a new line of "early born" humans living along the shores of Lake Turkana who were developing an unparalleled ability to adapt to the world around them. Something different was afoot: a species that was becoming, to borrow the phrase of Jacob Bronowski, "not a figure in the landscape, but a shaper of the landscape."

Improvements, however, have a way of creating new challenges. Now that a series of surprising and unintended events had taken human evolution into entirely new territory, our ancestors found themselves caught in a strange, runaway feedback loop that would favor the arrival of increasingly large-brained, increasingly intelligent, and increasingly helpless babies. All good, you would think. Except that every one of them would require more care and long periods of time to grow up. That would shake the social lives of Africa's gracile primates to the core and lead to yet another profoundly important twist in our evolutionary story.

CHAPTER FOUR

TANGLED WEBS—
THE MORAL PRIMATE

Morality, like art, means drawing a line someplace.
—Oscar Wilde

IN 2005 ENGLAND found itself mesmerized by the gruesome murder of fifty-seven-year-old businessman Kenneth Iddon. Each Sunday, Mr. Iddon would drive to nearby Deanwood Golf Club to play snooker with his friends, then return home around midnight. On February 1, 2004, before he could get out of the car in his driveway, prosecutors said three men bludgeoned him, dragged him into his garage, repeatedly stabbed him and finally killed him by severing his carotid. It all happened while his wife and stepson were in the family house nearby. Others in the suburban neighborhood later reported they heard cries for help, yet Lynda, Mr. Iddon's wife, and Lee Shergold, her thirty-one-year-old son by a previous marriage, said they never heard a thing.

They denied hearing any cries for help, the local prosecutor charged, because both Mrs. Iddon and her son had hired the three men who killed Mr. Iddon. They wanted his money, they said, all of it, not simply what Lynda Iddon might get in a divorce settlement. The irony was that when Mr. Iddon's will was read, he had left nothing to his wife. His entire fortune was bequeathed to his twenty-two-year-old daughter, Gemma. Neither Lynda, nor Lee, inherited a dime.

It's an old human story. Greed, hatred, envy, and violence. We have, in case you haven't read today's newspaper, been known to commit acts we call immoral. We find them abhorrent and disturbing, yet, given our immense numbers, we actually show our ugly sides relatively little. One of the reasons we call attention to the terrible things we do, and are horrified by them, is because the majority of us don't do them. We are the only animals that even struggle with the idea of morality because we are the only truly ethical animal.

Our moral tendencies are apparently so thoroughly wired into our psyche that they even reveal themselves in young children. For years psychologists—from Sigmund Freud to Jean Piaget to Lawrence Kohlberg—denied that infants or toddlers could have any sense of right or wrong. The traditional view has long been that babies arrive without the slimmest grasp of empathy, fairness, or other similarly moral sentiments. But recent experiments show otherwise.

At Yale University, psychologists Paul Bloom, Karen Wynn, and Kiley Hamlin placed infants between the ages of five and twelve months in front of a simple morality play where three puppets were throwing a ball. As the babies watched, one puppet rolled the ball to another puppet on the right, then that puppet promptly rolled it back. Next the center puppet rolled the ball to a third puppet on the left, who took the ball and, instead of rolling it back, ran off with it.

The infant audiences didn't cotton to this sort of behavior. Later when they were presented with the two puppets to which the balls were rolled, each with a pile of treats, and were asked to take a treat from one of them. Invariably they took a treat from the "naughty" puppet that had absconded with the ball. One one-year-old went so far as to smack the offending puppet on the head, raising the question, is violence a proper response to an immoral act?!

The primal depths to which our sense of morality runs, and the murkiness of what we consider to be a moral or an immoral act, were brilliantly illuminated in a thought experiment that British philosopher Philippa Foot conceived over thirty years ago. (American philosophers Judith Jarvis Thomson, Peter Unger, and Frances Kamm later expanded upon Foot's experiment.) It asks that you imagine you are standing on a bridge overlooking tracks down which an out-of-control train is hurtling. As your eyes follow the route of the train, you are horrified to find five people tied to the track. But, you are told, you can save the five doomed people if you flip a switch that will direct the train toward a second fork. The only problem is that there is also another single person bound to *those* tracks.

Now what do you do?

The vast majority of those who take this test, or variations of it, don't hesitate to say they would flip the switch and sacrifice one person to save five. It's not a perfect situation, but at least, the thinking usually goes, five people are being saved even if one must be sacrificed to spare them. (What would you do?) Several years later, Judith Jarvis Thomson raised the experiment's stakes by offering an alternative scenario. This time the train is headed toward our doomed fivesome, but the only way you can save them is by throwing a heavy object in front of the oncoming train. As it happens a large person is standing with you on the bridge. Should you push the person over the railing to save the same five people? Could you?

Deciding what to do in that second situation turns out to be a lot more difficult than in the original scenario. But why? The outcome is precisely the same: one human life sacrificed to save five. But this time it's personal. It's one thing to flip a switch; the logic is obvious and you can act by remote control. But it's quite another thing to look a fellow human in the eyes and then personally push him to his death below. Not that most people think all of these issues through coolly and logically before they answer. The reaction is visceral, primal.

If we see the rudiments of morality in examples such as these, it's not difficult to imagine how primeval versions developed among the tribes of early humans who had begun a million years ago to make their way out of Africa and take the human species, for the first time, global.

In many ways their situation was similar to another classic thought experiment that emerged in the 1950s out of what computer scientists call game theory. The problem is called the Prisoner's Dilemma and is based on the work of two mathematicians, Merrill Flood and Melvin Dresher, at the RAND Corporation. (Much later Albert W. Tucker added some formal touches to the game.) As much as we might like to think that our sense of fair play traces its roots to human kindness and altruism, game theory illustrates that deep down even the best behaviors stand on a practical foundation, a form of enlightened self-interest, at best. For our purposes the game goes like this.

Jack and Joe are arrested by the police and charged with robbing a bank. It's a given that both men are pretty reprehensible and more concerned about their personal freedom than they are about one another. The problem is the authorities don't have sufficient evidence to convict either one. So they separate them and offer each an identical deal: You can testify against your partner, and if he doesn't testify against you, you'll go free and he'll go to jail for ten years. If you both remain silent, then you'll each serve a short sentence. If you both testify against one another, you will each serve five years. And if you refuse to testify at all, but your partner testifies against you, you will serve the full sentence. (It's a fair bet, by the way, that the British police investigating the Iddon murder used exactly this strategy with the murderers, several of whom did eventually turn on their fellow conspirators.)

Scientists have found that if the game is played once, six players out of ten choose to testify against their partner. We shouldn't be too surprised that most people will rat out their partner because by testifying this one time, the best thing that happens is you walk away. The worst is that you end up with a split sentence.

If the game is played again and again, however, and the players can exact revenge on one another or reward good behavior, which is more the way life is, then the players gain enough feedback that they learn how their counterparts behave, and in time something interesting happens. Each player begins to cooperate with the other because each realizes that watching out only for himself (and choosing to turn in his partner) may result in his partner's punishing him the next round of the game. What happens? They both begin to choose to *not* testify, which results in both getting off with a slap on the wrist. A sort of morality emerges. Players begin to realize that if they treat others as they would like to be treated—the Golden Rule—life won't be perfect, but it will, on balance, be pretty good.

What all of this clearly reveals is that we are, and have been for some time, moral animals. But where does our morality come from? Why did it even evolve? Other animals (with the exception of some of our cousin primates) don't struggle with morality. Why should we?

The reason is because we are so shamelessly social.

By the end of 2011, Facebook, the current darling of the Internet, claimed 750 million active subscribers, who together pass an average of 700 billion minutes a month digitally engaged with one another. Since its emergence in 1993, the World Wide Web has rocketed from zero Web sites to 45 million, and counting. Last year, uncounted billions of us worldwide were busily talking, incessantly texting, or otherwise interacting with one another on more than five billion cell phones. These statistics aren't simply impressive examples of our ingenuity; they represent monuments to our primal need to connect with one another.

The convivial interactions of ants, termites, and certain kinds of algae not withstanding, we humans are indisputably the most socially complex species to have ever emerged on planet earth. We cannot bear to be disconnected. For a human, the worst kind of torture is solitary confinement, a punishment that can lead to depression, hallucinations, and madness. We can't seem to help keeping constantly in touch, literally or metaphorically, always reaching out, laughing, crying, gossiping; talking at, with, or about; watching, gaping, glancing; listening in, listening to. Even in ignoring one another we are tacitly bonding by acknowledging that others are around us to turn our nose up at. Ugly as they are, hatred, jealousy, envy, rage, discrimination, even murder, could not exist if we weren't, first and above all, bound to one another inextricably. It is possible, I suppose, that in some parallel universe each of us could be like Dickens's Ebenezer Scrooge, "as solitary as an oyster," but if that was the case, not only would love, marriage, business, and cities be out of the question, so would Super Bowls, World Cups, global trade, finance, symphonies, and the rest of human civilization with them. We have built the world that we have built, either in cooperation or competition with one another, but *we* have built it.

The currency of our connectedness is communication, which is so mystifying that legions of scientists still labor ceaselessly to unravel its complexities. We communicate using language, but also by tapping uncounted libraries of nonverbal behaviors, too—laughter, tears, body language, and facial expressions, not to mention painting, mathematics, sculpture, music, and dance in all of their variability across all the cultures of today, yesterday, and futures to come. Each of them represents a handful of the unending inventions we press into service to express to one another what we are thinking, feeling, exploring, want or wish for, fear, hate, and love.

All of life is linked, from amoebic protozoa to the invisible oxygen-fixing microbes that make the enormity of sequoia trees possible. We aren't alone in that. Of the one hundred quadrillion cells that each of us carries through the day, for example, only 10 percent belong to us. The rest are outsiders, the microbial flora and fauna that live in our stomachs and organs and dine out on the surfaces of our bodies. Yet without those trillions of hardworking microbial committees, not a single one of us could make it through the day. We need one another.

Global ecosystems likewise require connection and communication. Symbiosis and competition make the world, quite literally, go around, because without the interactions of life, from the oceans' microscopic phytoplankton to rumbling migrations of wildebeests, our world might just as easily be as dead, phlegmatic, and torridly hot as Venus, or as cold and parched as Mars. In short, life and communication can't be separated. But in our case the affliction is unusually deep and complex.

We can trace these tendencies back to the early mammals that began to gather evolutionary momentum after the dinosaurs were wiped out sixty-five million years ago. A cerebral innovation that mammals brought into the world was the brain's limbic system, the seat of our emotions. Lots of mammals are social and live in packs, prides, and herds, but our particular primate line traces its roots to mammals that evolved into monkeylike creatures that mostly stuck to the jungle, but also eventually ranged out into the savanna, where they lived in small clusters.

From these, and in a relative blink, multiple species of primates arose between twenty-five and five million years ago. Since every other human species that has ever managed to make its way into existence is now extinct, except for us, our closest living primate relatives are chimpanzees, bonobos, and gorillas, all of them exceedingly sociable. This tells us that when our earliest human ancestors found themselves stranded on Africa's expanding savannas, they were already inveterately communal. After all, shortly beforehand they all shared a common ancestor.

The best we can figure, they lived in troops of twenty to fifty, possibly more, sometimes less, traveled together, shared food, fears, sex, and other dangers and amusements. So tight were their communities that there would have been zero chance of any member of the troop running into another and not recognizing him or her immediately. (Maybe this is why studies reveal we have trouble handling more than twelve to fifteen truly personal relationships, Facebook not-

withstanding.) The only strangers these creatures would have encountered would have hailed from other troops or even other species,
and those encounters would likely have been as strange as you or me
running into a Sioux warrior from 1825 at the local mall.

As our ancestors were left to wonder their new, more dangerous
grassland environment, the ties that bound them would have grown
tighter than ever. The aphorisms "misery loves company" and "there
is safety in numbers" might arguably have found their origins here.
On the savanna there were more predators, but fewer places to hide,
less food, less water, than the jungle provided, even more competition
from other troops given the dearth of resources in their new home.
Disease and injury surely did in more than one man, woman, or child
as the troop wandered from place to place among the volcanoes, and
along the shores of lakes like Turkana, every loss threatening to rub
the clan down to a size that made it impossible to survive the next
disease or environmental blow.

Now add to all of this the pressures early-born children brought to
the mix. You find yourself awakening every day bound to your fellow
creatures working to survive, raising children, forging friendships and
alliances, scrounging for food, and communicating as much as your
brain and body will allow. You have no other choice because if you
don't, you will die. But (there is always a but) living in such a tight
community also means competing with the selfsame creatures you
rely upon for mates and status and resources. Tricky situation, because
it requires balancing what you want for yourself with everyone else's
needs. It means simultaneously taking care of number one *and* watching out for those around you. This is the central paradox of the human
condition—balancing, constantly, two seemingly opposite needs.

We see the evidence of this continuing dilemma every day from sibling rivalries to office politics, from international trade to military bal-
ances of power. Every day's headlines and news reports are dramatic
testaments to our struggles to act morally. Robbery, terrorism, murder,
heroism, stock-market crashes, war, charity, law, international aid,
trade, and political intrigue are all examples of our attempts, and fail-
ures, to deal fairly and ethically with one another, writ large. For the
bands of our predecessors struggling to survive on Africa's plains, however, this was new territory, and it required the development of some
kind of moral code.

Think back for a minute on the Prisoner's Dilemma. Existence on

the savanna among small troops of hominins would have been remarkably similar to Jack and Joe's situation after their arrest. If you were a
member of the troop, it would make no sense to repeatedly abuse
those around you even if you had the power to do it. If you did, you
would quickly find yourself persona non grata, shunned by the troop,
or, worse, dead.[1]

On the other hand, what about *your* needs? You ignore those at your
peril, too. You require a mate, food, safety, and personal power just as
much as anyone else. To deny these could result in your death, too. A
dicey dilemma.

We are clearly and painfully still struggling with these issues, but
over the long haul evolutionary forces encouraged our ancestors to
cooperate enough with one another that they managed to make it to
the twenty-first century. Like Jack and Joe, experience taught our
ancestors that on balance cooperators tended to stay alive long enough
to have babies and pass their genes along. Cooperators wouldn't have
always gotten everything their way, but they also wouldn't have been
tossed out of the group to fend for themselves, a sure death sentence
given the harsh realities of life a million years ago. Success in such a
tight community depended increasingly upon how deftly you navi-
gated and balanced your relationships with your peers. Accomplishing that, however, required an even more powerful brain than the
one nature had already bestowed on struggling predecessors.

In the 1990s a Liverpudlian psychologist named Robin Dunbar conducted research that illustrated a correlation between the size of an
ape's brain and the size of the troop in which he lived. The larger the
group, the larger the brain. He argued that bigger troops drove the
evolution of larger brains because every new addition to the group
ratcheted up the number of direct and indirect relationships each member had to keep track of. Juggling more relationships required a corre-
sponding boost in intelligence. Evolution would have favored smarter,
larger-brained members of the troop because they would have been
better equipped to track the growing social relationships between fellow primates.[2]

Something similar was happening among our direct ancestors on
the plains of Africa a million years ago, with an important additional
ingredient. The driving force behind the evolutionary change wasn't

merely the size of the group; it was the complexity of the relationships inside it. Our ancestors were much smarter than Dunbar's primates, and the dynamics of their relationships would have been more complicated. After all, at that time they were the smartest creatures on earth. Greater intelligence is a multiplier of complexity because it increases the number of *factors* in relationships. It adds more variability, more motives, more intrigue and nuance, and, in turn, drives up the advantages of possessing the additional neuronal firepower needed to constantly calibrate exactly why people are acting the way they are, and more particularly why they are acting toward *you* the way they are.

Human relationships are dynamic and fluid. They change constantly. Rarely do we unquestionably love, or entirely distrust, the people in our lives. Mostly our relationships slide along a continuum in a never-ending exchange of interpersonal, emotional, and mental calculations. The social lives of our ancestors may not have reached the Machiavellian proportions of the Soviet politburo, the court intrigues of Henry VIII, or even the office politics of *Mad Men*, but, generation by generation, you can be sure they were getting increasingly complicated. And that would have required the introduction of a new and powerful behavior: deception. Or more precisely, as you will see, our ability to detect deception.

At this point in the evolution of life on earth, deception was clearly far from new. Prevarication is an essential part of existence and has been for far longer than our kind has been around. Venus flytraps pose as beautiful flowers to lure their quarry to their doom. A leopard's spots or a chameleon's changing colors dupe prey and predator alike. Young spider monkeys have been known to fake predator calls so they can scatter their elders who are dining on recently found food, then pilfer the goods before others in the troop are any the wiser. The cake for natural deception might have to go to a particular shallow-water anglerfish (there are many species) that looks remarkably similar to a rock encrusted with sponges and algae. At the end of its head extends a thin spine that supports a piece of itself that would be the envy of every avid reader of *Field & Stream* magazine. It looks exactly like a small living creature right down to the pigment along its flanks and "eyes" at the top of its faux head. The anglerfish even wiggles the bait so that it seems to be swimming along just like any number of other fish in the sea. When a hungry fellow fish arrives to take the bait, the angler gulps it down before it has even realized it is the hunted, and not the hunter.

There is, however, a difference between these deceptions and the human variety. The human sort that was shaping up a million years ago was conscious, which is to say planned and driven not purely by genetics. In these ancestors we begin to see the evolution of chicanery in the service of self-interest at a level never before seen, the deliberate, premeditated variety.

In some ways cheating of this sort was inevitable. It is the flip side of the primal moral code that was evolving at the same time. As early humans found ways to cooperate and trust one another—which was absolutely necessary if they hoped to survive—wasn't it equally inevitable that deception would also emerge? It was, after all, a powerful way to serve personal ends without having to deal with the overt danger of direct confrontation inside the troop—a perfectly understandable, even brilliant, adaptation when you consider the circumstances. Deception was an accommodation, a kind of compromise, except that only one party was in on the secret. If you can cheat and get away with it, you're riding on the backs of others to your benefit (and their detriment) without anyone's knowing it or becoming even the slightest bit upset about it. Not a bad ploy, if you can get away with it.

Of course over the long haul getting away with it would have to fail, because if it succeeded indefinitely, the spread of bad behavior would unravel the success that sustained the group, something like the way a too-successful parasite will kill off its host (and itself if it succeeds). Ultimately, the bad behavior has to stop, or at least be controlled. If, among a small band of *Homo ergaster*, for example, food was stolen, personal hoarding got out of hand, slackers consistently failed to pull their weight, or mates continually cheated on one another and refused to protect and care for their families, the group's social fabric, and the trust that kept it woven, would fly apart. No one would win.

So in the arms race of ever-improving minds, detecting bad behavior would have been an extremely important skill for our ancestors to develop—an antidote to deception. And it turns out they did, at least according to evolutionary psychologists Elsa Ermer, Leda Cosmides, and John Tooby.

We all engage in what scientists call social exchange. We agree to do something for someone in exchange for his or her doing something in return for us, either now or in the future. We do this because on some level we believe that the exchange works to our benefit. So does the other person. "You scratch my back, and I'll scratch yours." Everything from family relationships to world economies rest on this fundamental human behavior. And for our ancestors, it would have been essential to their common survival.

But what happens when you scratch someone's back and he or she doesn't scratch your back back? According to tests Ermer, Cosmides, and Tooby conducted with everyone from hunter-gathers in the Amazon to university students in Europe, Asia, and the United States, we humans have unerring radar for sniffing out those who cheat the system; a kind of social immune system that finds and exposes freeloaders. Not that this radar is perfect in all matters of deception. The tests indicated we are not all that skillful at unmasking trickery, infidelity, or accidental cheating, but when it comes to the scratch-my-back-and-I'll-scratch-yours variety, we are extraordinarily talented.

Uncovering any physical evidence of this special ability to expose cheaters among the dust and bones of our long-lost predecessors is, unfortunately, impossible. There are, to paleoanthropologists' everlasting sorrow, no fossils of behaviors. But in another study, cognitive scientist Valerie Stone at the University of Denver did find a different kind of physical evidence, this within the human brain, which indicates our ability to suss out social-exchange cheaters is wired somehow into the wetware between our ears, a little like the ability to learn language.[3]

At the heart of Stone's investigation is R.M., a man who had in a bicycle accident damaged a rare combination of areas in his brain—his orbitofrontal cortex, temporal pole, and amygdala. R.M.'s accident was tragic for him, but fortuitous for science because all three of these areas are crucial to social intelligence, particularly in making inferences about others' thoughts or feelings based on, say, an angry tone of voice, a scowl a smile, or a person's body language.

Stone devised a test for R.M. to see if particular kinds of if-this-then-that statements were more difficult for him to understand than other kinds. She asked him to analyze three different types. One, for example, dealt with precautions. "If you work with toxic chemicals, you have to wear a safety mask." Others involved descriptive rules. "If a person suffers from arthritis, then that person must be over forty years old." A third kind of problem dealt with social, scratch-my-back-I'll-scratch-yours contracts. "Before you go canoeing on the lake, you first have to clean your bunkhouse."

R.M. had a difficult time correctly answering the social-contract questions, like the one about the bunkhouse. The difference between getting those correct compared with correctly answering the precautionary questions ("If you work with toxic chemicals, you have to wear a safety mask") was a whopping 31 percentage points.

Stone concluded that uncovering cheaters was so crucial to survival that evolution favored neural wiring optimized for understanding when someone was not living up to his or her promises. As luck would have it, R.M. had injured exactly the parts of the brain involved in this wiring.[4]

You might think if we were this good at spotting cheaters, we would be equally talented at detecting other kinds of deceptions. But that doesn't seem to be the case. A few years ago, two psychologists, Charles Bond and Bella DePaulo, wondered exactly how sharp we were when it came to catching others in the act of pulling the wool over our eyes. Rather than conduct their own study, they organized a study of studies, analyzing documents from 206 other research projects focused on various sorts of human deception and our ability to discover it. They pored over no fewer than 4,435 individuals' attempts to dupe 24,483 others and found that the dupers were unmasked by the dupees only 54 percent of the time, or just a little better than you or I would do if we flipped a coin.

It turns out that one of the reasons we aren't better at spotting lies is because we have learned to be almost (but not quite) as good at hiding the truth from one another as we are at uncovering it. It's not that we are horrendously inept at calling out the equivocators among us; we have just learned to improve our lying and fakery. In the ongoing arms race between deceivers and truth seekers, the competition is so close that it's resulted in a kind of Mexican standoff. According to the research of one of the true pioneers in the field of kinesthetics, or body language, psychologist Paul Ekman, this has resulted in several intriguing insights about the way we behave in one another's presence.

Sigmund Freud famously wrote in 1905, "No mortal can keep a secret. If his lips are silent, he chatters with his fingertips; betrayal oozes out of him at every pore." Ekman and his research collaborators found that the great Austrian psychoanalyst was right, our bodies can often subvert our best attempts to deceive, but not in the most obvious ways and rarely in the ways we read about so often in popular magazines. For example, we are outstandingly skilled at hiding the truth verbally, a little less good at hiding it by controlling our facial expressions and hands, and least effective of all at hiding the ways our legs and feet can reveal our fabrications. The parts of us over which we have the most conscious control are the parts we've become particularly good at masking.

Bond and DePaulo theorize plenty of additional reasons for why our rate of sniffing out deception is hardly better than our ability to accurately predict a coin toss. For one thing, we are fundamentally trusting creatures, predisposed to believe those we deal with because it's rare that our dealings with them result in a catastrophic or dangerous lie. (If that were the case, we would all be far more paranoid, which would create its own set of unsavory difficulties.) Many of those we spend most of our time with tell us plenty of harmless fibs. How good we look that day, for example, or how funny a joke is; that they were late for a meeting because they had trouble starting their car, or the dog ate the weekly report—that sort of thing. Even if we don't believe everything we hear (or are pretty sure that others don't believe everything we say), this variety of truth bending isn't damaging, and sometimes it's even constructive. So our tendency to miss untruths might also be a matter of motivation because we aren't generally dealing with a world-class con artist who is hiding a dangerous whopper that puts our lives on the line. Every day is filled with rationalizations, self-deceptions, white lies, and all varieties of other spin.

The competition that required liars to outfox their dupes, and then dupes to figure out the deception strategies of good liars, and so on, almost certainly contributes to one of the neatest tricks the human mind is capable of—imagining it is not the mind it is, but someone else's.

If you happened long ago to be engaged in either side of this liar-dupe battle as our ancestors surely were, one of the best weapons you could possible devise would be the ability to shift your point of view and imagine yourself in the shoes of the person who might be lying to you (or the shoes of the person you are trying to deceive). This ability would allow you to not only imagine the situation from the other person's viewpoint, but you could look at yourself from the outside and, perhaps, spot flaws in your own dissembling techniques. This is the psychological equivalent of placing two mirrors face-to-face, creating an escalating infinity of images, except in this case you can create an infinity of viewpoints that shift back and forth reacting one to the other. (This recursive ability turns out to be crucial to human consciousness, as we will later see.)

Novelist and screenwriter William Goldman beautifully illustrated this contest when he wrote a scene for his charming, comic send-up of the classic fairy tale, *The Princess Bride*. A Machiavellian (and hunchbacked) master of deception and intrigue named Vizzini agrees to face off with the book's masked, Robin Hood–esque hero in a battle of wits. At stake is the book's beautiful, but flinty, heroine. The two men sit, each with a goblet of wine in front of him. One of the two goblets is deadly, tainted with a poison called iocane. Under the rules of the battle, the masked hero already knows which goblet is poisoned because he put the iocane in it, but Vizzini alone gets to chose which goblet they each must drink from. If he can calculate which goblet has the poison, he will chose not to drink it, killing his rival. The scene unfolds like this . . .

"Your guess," he [the masked hero] said. "Where is the poison?"

"*Guess?*" Vizzini cried. "I don't guess. I think. I ponder. I deduce. Then I decide. But I never guess."

"The battle of wits has begun," said the man in black. "It ends when you decide and we drink and we see who is right and who is dead." . . .

"It is all so simple," said the hunchback. "All I have to do is deduce, from what I know of you, the way your mind works. Are you the kind of man who would put the wine in his own glass, or the glass of his enemy?"

Vizzini then goes on to logically slice and dice the situation, not to mention the psychological makeup of his nemesis, with each deduction making the hero increasingly nervous until at last Vizzini comes to his conclusion.

"I have already learned everything from you," said the Sicilian. "I know where the poison is."

"Only a genius could have deduced as much."

"How fortunate for me that I happen to be one," says the hunchback, growing more and more amused . . . "Never go against a Sicilian when death is on the line."

He was quite cheery until the iocane powder took effect.

The man in black stepped quickly over the corpse.

How did our masked hero win the battle of wits? How could he be
so sure his deception wouldn't be found out, leaving him to drink the
poisoned wine? Here's how: He had spent two years building up an
immunity to iocane. It didn't matter which goblet Vizzini drank from.
Both were poisoned! And with that move, he raised the evolutionary
stakes. Unfortunately for Vizzini, he fell one step behind in the arms
race and was selected out.

Goldman's little story encapsulates the ongoing battle our ancestors
found themselves fighting. In dealing with their increasingly complicated relationships, those early humans who became more skilled at
climbing inside the minds of the others around them would more often
win the battle of wits. They would also enjoy a decided evolutionary
edge because they would excel at practicing deception as well. This
makes us a tricky species indeed.

Psychologists call this unique human ability to hop back and forth
between our own point of view and someone else's Theory of Mind
or ToM. Uncovering deceit isn't the only time we use it (though it's
certainly a helpful application). We employ ToM almost every waking
moment we are interacting with others or thinking about interacting
with them. It is, if you examine it closely enough, the foundation upon
which all human social commerce is built. It enables us to empathize,
anticipate, and outfox. We exercise it when we talk with one another,
or about one another. It's in play when we lie awake in bed wondering why our spouse or girlfriend or boyfriend did this or said that, or
the boss gave us a hard time about the quarterly report, or even why
he put his arm around our shoulder and said, "Atkinson, helluva job!"
What, we wonder, did he really mean by that? To put it bluntly, we
are a species incessantly thinking about what everyone else around us
is thinking.

Yet ToM has even broader applications and effect because it provides us with a remarkable talent for running infinite numbers of
what-if scenarios, all simply by firing the neurons in our own heads.
You can imagine one of our ancestors wondering, "What if the leopard

jumps out of that tree? What if I get caught wooing this female? What if I come back with some meat and give it to Woog? Will that get me a little troop cred? Is it worth the trouble?"

What-iffing allows you to not only step into someone else's place, it gives you the magical power to step into the future and prepare for what might happen next. Or to create parallel universes where you can run multiple scenarios about taking this action or that one, then weighing them to see which might lead to a better outcome; something we call, among other things, imagination. As I write these words, my mind is what-iffing furiously about which are the best scenarios to run by you to make the points I want to make.

Being able to say to ourselves, "If this, then that," builds the infra-structure of human creativity (more on this a bit later in the book). Scenario building is pure make-believe, a return to that time in our childhoods when we used to say, "Let's pretend . . ." It is a way to create and explore possibilities that don't exist in the real world, but live completely in the universe of our minds, and nowhere else. A remarkable thing.

Misfired Mind Reading

Mind reading, and the abilities that make it possible, can also mis-fire. (A lot of evolutionary innovations do.) It can make us chronic worriers, stuck in endless loops of stomach-churning scenario building, erecting realities that aren't real at all while we suffer through them as if they were; percolating endlessly on this or that possibility and applying it to bosses, significant others, children, and just about every decision we make. We may be the proud and mighty scenario-building animal, but we also invented nail-biting, hand-wringing, and acid reflux. Sometimes imagining what someone else thinks can be absolutely paralyzing—how your mother, or the vicar, or even another version of yourself might view your first sexual encounter, for example.

Whatever the uses to which we personally put the mind-reading/scenario-building powers that our ancestors developed, this much is beyond debate: no brain in nature had ever before seen its like. This is, neurologically speaking, inconceivably difficult to pull off. It demands billions of neurons and requires that the newest *and* most ancient parts of the brain be wired deeply to one another. Valerie Stone's experiments with R.M. illustrated this. R.M., remember, had damaged his amygdala, whose evolutionary roots are reptilian; the temporal pole, which is part of the limbic/mammalian brain; and the orbitofrontal cortex, which is among the newest cerebral additions to have evolved. Our ancestors were becoming chimeras, of sorts, creatures built out of the spare parts of both ancient and modern evolutionary mutations, an amalgamated animal, both ancient and new, self-aware, yet driven by unconscious, subterranean impulses. In a phrase, we were becoming really complicated.

We can't know for certain when the rudimentary ability to climb inside another's mind evolved. Such abilities are almost certainly not the result of a lone adaptation. More likely they resulted from scores of suites of adaptations that surely took an immense amount of time to emerge. One point two million years ago the robust human lines had seen their last days. It was a good run, but the evolutionary path followed by the gracile apes, unlikely as its success was, had won out. Yet, who would have predicted it? Not even a what-iffing creature. Larger brains forced earlier births, earlier births lengthened and complicated childhoods that created minds increasingly shaped by personal experience, which in turn made the mind more creative and adaptable. Brains over brawn.

And as if this wasn't messy enough, now longer childhoods were producing people that were genetically similar, but behaviorally unique; every troop was loaded with highly complicated individuals, each with her own talents, psychological baggage, foibles, and agendas. Yet they bonded, despite their individual needs and selfish competitions. An odd, astonishing species, or group of species, if ever there was one.

A mix this complicated would still seem doomed to failure. How do you weigh and balance all of these competing needs; manage the increasing complexity of your own motives let alone the motives of those around you while avoiding simultaneously alienating the allies you need? Depending on the situation, did "might make right," was it better to be conciliatory, or was deception the best path?

It all had to be worked out, and apparently it was, otherwise you and I would not be here. Out of this complexity, these competing needs, a moral ape was born, made possible by the early childhood that had shaped our gracile ancestors. They had managed to find strength in numbers, and a workable code of conduct. It may not have been perfect, but they were successful enough that they had begun to take the species, several species actually, global. They had not only become a moral ape, they had evolved with an irrepressible case of wanderlust.

CHAPTER FIVE

THE EVERYWHERE APE

*Now my own suspicion is that the Universe is not only queerer
than we suppose, but queerer than we can suppose.*
—J. B. S. Haldane

OUR SPECIES IS the most itinerant and restless animal on the planet. That's a simple fact. You will find polar bears on the ice sheets of the Arctic, silverback gorillas in the mountains of central Africa, reindeer in northern Europe, tigers in India, and penguins in the Antarctic, but you will find humans in all of those places and more. We are the only mammal that inhabits all seven continents, and it doesn't matter to us how hot, how high, how humid, or how frigid the geographics are in which we live. We have even found our way, God knows how, to thousands of remote islands around the planet that amount to no more than an oceanbound fleck of dirt that your eye could easily lose looking at a decent-size map—Easter Island, for example, whose nearest inhabited neighbor is more than a thousand water-soaked miles away. We are everywhere. But we weren't always so. At one time we were almost nowhere. How we went from a few locations to many makes a fascinating story. It also says a lot about who we are.

At the tip of South Africa where the Indian and Atlantic Oceans meet lie shores of basalt rock that look out on an expanse of cold and turbulent water that doesn't see another shoreline until it meets the ice cliffs of the Antarctic more than a thousand bracing, windblown miles away. If ever there was a place you could call the end of the earth, this is it.

Seventy thousand years ago, a few hundred human beings lived here; anatomically modern humans or AMH, as anthropologists like to call them. They were like us in every way it seems, except for the technologies they used to survive. They were bereft of cell phones and SUVs but looked like us and carried around the same evolutionary and psychological baggage we do. In those days, they were also the last remaining members of our species, a tiny enclave of humanity twisting precariously at the end of an evolutionary thread, rubbing elbows with extinction.

One hundred and twenty thousand years earlier this species, one that would later name itself *Homo sapiens*, had come into existence, a new branch of the human family, split off from an earlier primate that had arisen on the Horn of Africa, where so many other varieties of humans had emerged.

This particular tribe, the one that lived along Africa's southern shore, were gracile, built for running, and clever hunters. Because of their high foreheads, prominent chins, and brains weighing more than three pounds, triple the size of those of the first upright walking primates from which they had descended, they looked far less apelike than their predecessors, though you could certainly see the family resemblance. They were inventive, too. Not only did they use fire, they controlled it, cooking food with it and applying it like a tool to harden and shape an impressive assortment of other cleverly fashioned gadgets—knives and axes more advanced than any used before. They may have been at the ends of the earth, but this was the Silicon Valley of its time, a hotbed of innovation. They had also developed an extremely powerful way to communicate—words.

Fortunately for these last survivors, the land was Eden-like. Not tropical, but temperate and sustaining. What it lacked in the big game that walked the northern savannas, it made up for with lush stores of fruit, nuts, and beans, and an inexhaustible supply of protein-rich seafood. Life must have looked very good. After all, deprived of CNN and the Weather Channel, they had no way of knowing they were the last representatives of their species, nor that much of the world and the continent beyond their small slice of paradise had been under climatological assault for thousands of years. A harsh and unrelenting ice age had already wiped out others like them farther north. Europe, Asia, North America, and the Mediterranean had been buried for millennia beneath uncounted miles of snow, howling winds, and frozen seas. Oceans of water were now locked in enormous ice sheets, leaving seas more than 225 feet shallower, and the rest of Africa chilled and bone-dry. This was the apocalypse.

It was possible that they were not entirely alone. Tiny pockets of other modern humans may have survived the ice epoch in the north and west of Africa, but no one can say with certainty.

No, this was probably it. Just a few hundred people dug in, the current crop of an extended family who had colonized the area as many as fourteen thousand years earlier. One catastrophic event, a plague, a typhoon, or a freeze, and that would have been the end of Homo sapiens. And none of the seven billion of us who exist today would ever have been the wiser; in fact we would not have "been" at all. We came that close to being snuffed out.

That, at least, is how paleoanthropologist Curtis Marean sees it. It's a sobering thought, the idea that we were closer to extinction than today's mountain gorillas, and not much better off than India's dwindling prides of tigers.[1]

Plenty of scientists dispute Marean's scenario. It wouldn't be paleoanthropology if they didn't. Our past is a messy business, and today's efforts to understand how we came into existence, based largely on the ossified leavings found in the world's dust and rock, has been something like a blind man's trying to describe the details of a football stadium by feeling his way through it. If we didn't have ourselves around to inspect, we would know more about Homo habilis and Neanderthals than we know about Homo sapiens. You would think that our being among the most recently arrived branch of the human family we would be knee deep in the evidence of our own existence, but that's not the case. Outside of Africa, fossils of early Homo sapiens are nearly nonexistent. Thankfully, we have been learning to read the path of our evolution in our DNA (see sidebar "Genetic Time Machines," page 76), and that, together with some meager findings in the fossil record, has illuminated the story of our emergence at least a bit. The story goes something like this.

Between 160,000 and 200,000 years ago the first anatomically modern humans emerged, probably near Ethiopia. (But there is anything but universal agreement on this.) Among these was a woman, now called the matrilineal "Eve," the "mother" of the human race, though that term is a little misleading. Eve wasn't herself the first modern human, and unlike the biblical Eve, she wasn't the only woman alive two hundred thousand years ago. She was, however, the sole woman alive then that still has descendants today. Other modern human women

Genetic Time Machines

When it comes to DNA, the only certainty is change. It's restless. As DNA alters, so do genes, and when genes mutate and unwittingly express new traits, their accumulated mistakes eventually result in entirely new species—by some estimates, thirty billion separate forms of life over the past 3.8 billion years. Despite the messy nature of genetic mutations, they create markers whose rates of change are startlingly predictable. These signposts enable scientists to calculate, with reasonable, but far from perfect, accuracy where in the evolutionary picture your particular branch of the family tree diverged from other branches.

Two primary kinds of DNA allow scientists to pull off this neat trick. One is the DNA of organelles that live within each of our cells, called mitochondria. Groups of mitochondria exist within each of the fifty trillion cells that make you and me possible. In an evolutionary partnership agreed to some two billion years ago, some single-celled bacteria took up residence in other single cells, but refused to give up their DNA in the bargain.* The relationship has remained unbroken ever since. Today, in exchange for the protection and nutrition they receive living within other cells, mitochondria create the chemical energy needed to power nearly every plant and animal on earth, including us.

The second kind of DNA is the nuclear variety, the sort that belongs directly to you and me and within whose cells those mitochondrial guests live.

It is now possible to take a fossil of our ancestors, closely scan the DNA trapped within (usually mitochondrial because there is more of that than the nuclear variety), and, if the information is robust enough, compare it with samples of our DNA and see how different the two are. Then by comparing the markers—the average rate of mutations over time—we can estimate how deep in the

---

*Life evolved soon after Earth itself came into existence some four billion years ago. The first cells were prokaryotic. The best guesses for the time when eukaryotes (cells with mitochondria) evolved range from just below 2.0 billion years to around 3.5 billion years before the present. The early fossil record for single-celled organisms, as you might expect, is sparse, so it's tough to set the exact date of this remarkable bargain.

past the two genomes were once identical and when they went their separate ways. This is a little bit like standing on the limb of a tree and pacing off the distance between the branch you are standing on and the one from which it sprouted. Each pace provides an indication of how long ago you and other tree limbs separated.

The ancestor that all humans share going back to *Sahelanthropus tchadensis* (see The Human Evolutionary Calendar, page 7) would be represented by the tree's trunk. Each divergence, each limb, represents a new human species—*Homo habilis*, *Ardipithecus ramidus*, *Homo rudolfensis*, and all the rest. Some lead to new branches others, some don't. You can also imagine the mutations themselves as the landscape through which a kind of time machine can travel with the genetic markers as mileposts that indicate how far back or forward in time you have journeyed.

Whichever metaphor you chose, this is how scientists can compare our DNA with a Neanderthal's and conclude that we parted ways from a common ancestor—*Homo heidelbergensis*—200,000 to 250,000 years ago. Or how they have come to discover that Neanderthals and Denisovans both shared a bed with ancestors of ours whose offspring eventually made their way to Europe, Asia, and New Guinea, even though, especially in the case of the Denisovans, we have almost no fossils to inspect.

A variation on this same technique (more often this time looking at nuclear DNA) makes it possible for scientists to track down the patterns and timing of our own global wanderings—when one group remained in central Africa, for example, but another headed north. When some members of that tribe made west into Europe and others branched off to Asia and the east. This is because our DNA has mutated as we have traveled the world, though not enough in the past 190,000 years to have sprouted an entirely new species. These mutations indicate where we lived, and when.

lived during her time and before it, but she is the one to whom every living human today is related. So it's more accurate to say she is our "most recent common ancestor," at least when looking at mitochondrial DNA as a marker.

Thanks to the genetic records all creatures carry within them, and thanks to the ability of computers to compare them, we are developing a clearer, if not pristine, picture of how much we have in common with our fellow humans, when we parted ways with them, and how we have, ourselves, managed to make our way from a couple of pockets in Africa to nearly every spit of land earth has to offer.

If Marean's theory is correct, the first "moderns" that arose in the Ethiopian plateaus must have spread out west and south during a population explosion shortly before the punishing ice age, known today by the memorable meteorological term Marine Isotope Stage 6 (MIS6), began to take its devastating toll. This climatic shift sabotaged life everywhere, as we will see, and may have been further boosted by the largest known volcanic eruption in the history of earth on Sumatra, Indonesia, which blasted ash into the stratosphere, causing a "volcanic winter" that rapidly accelerated the cooling of earth. (See "Killer Explosion?" sidebar, page 80.)

Other genetic studies indicate that sometime between one hundred thousand and eighty thousand years ago, three lines of *Homo sapiens* made off in separate directions from East Africa. One headed south and became the ancestors of today's Central African Pygmies as well as the Khoisan (Capoid) peoples of South Africa. A second genetic group migrated to West Africa, but also departed the continent by way of the Arabian Peninsula. Many West Africans are descended from this branch, and so are many African-Americans and South Americans who, millennia later, were transported across the Atlantic in slave ships. The third branch remained on the Horn of Africa, but others of them branched northwest and north. From these migrants descended the people who today live along the Nile Valley, the Sahara, and the Niger River, which flows through, of all places, Timbuktu in Mali into the Gulf of Guinea. Some of these people also found their way out of Africa. Ten percent of today's Middle Easterners have the blood of this third group running in their veins.[2]

Given these apparently enthusiastic migrations, you might think that as a species we were finally off and running, but there was that wintry climate that was setting in. By seventy thousand years ago it was in full, frigid swing and had begun to systematically rub out life everywhere on the planet. (We are living right now in what scientists call a slim "interglacial" period of this ice epoch, a bit of information that is itself chilling.)

Genetic studies confirm that during this time *Homo sapiens* underwent what scientists call a "bottleneck event." That is to say, we had been worn down to something like ten thousand total adult members, a troop or tribe here or there, scraping out a living, probably along ocean shorelines and receding lake beds.

Ice ages rarely result in cold weather in Africa. Instead they parch the land, turn rivers into dry wadis, evaporate lakes, and wipe out the sustenance each provides. During some of these periods, the Nile itself was reduced to swamp and muck. Even today the continent is filled with ancient lake beds scarred by desiccated mud cracks that testify to exactly how arid the landscape had become. Whichever humans survived the first waves of these droughts, they had tools, but little else, and when water disappeared, so did the other animals, nuts, tubers, and fruit that supported them. Being at the top of the food chain did them little good once the chain itself was demolished.[3]

Dramatic as the scenario is, it's unlikely that the small tribe at Marean's Pinnacle Point represented the very last bastion of Earth's *Homo sapiens*. More likely they were among dozens of tribes that the changing climate squeezed into small pockets throughout the continent. Each being winnowed down until they must have wondered daily how much longer they might make it. By this time early forms of trade had undoubtedly developed, but the increasing isolation would have made it more difficult to stay in touch, share resources, or help one another out.

Eventually, however, the climate relented. For three million years—an appalling length of time to us, yet less than one thousandth of the planet's life—Earth had been undergoing some of the most erratic climate fluctuations it had ever seen, shifting from cold to warm and dry and wet every few thousand years. To make matters worse, for three hundred thousand years Earth's orbit around the sun had been elongating. That led to even deeper and more frequent climatic swings.[4]

But finally, fifty thousand years ago, this particular climatic pendulum began to swing in the opposite direction, and just as the ice had once relentlessly crept from the polar caps to endanger the species at lower latitudes, it now casually reversed itself, and Africa grew warmer and wetter. The sparse pockets of the human family, like Marean's survivors living at the tip of Africa, and elsewhere here and there, again found themselves blossoming and fanning out. Isolated tribes, separated by heat and desert and their own reduced numbers,

Killer Explosion?

Seventy millennia in the past, long before the pharaohs of Egypt ruled the Nile, even three hundred centuries before the cave painters of Lascaux began doing their remarkable work, the most powerful volcanic explosion to rock the planet in two million years shattered the island of Sumatra, Indonesia, in an area now known as Lake Toba, and nearly wiped out every *Homo sapiens* on Earth. Or at least it may have. The explosion of rock, ash, and hot magma was so violent it's difficult to find the words to characterize its power. Scientists have coined multisyllabic terms like megacolossal and supereruptive. It was twice as powerful as the largest eruption in recorded human history, which took place in 1815 at Mount Tambora in Indonesia. Historians called the twelve months that followed it the "year without a summer" because the globe-circling debris from the eruption so severely cooled the planet.

It is precisely this kind of climatic effect that makes the Toba explosion so interesting. The evidence indicates that it spewed between twelve hundred and eighteen hundred cubic miles of the planet into the sky. Some scientists believe that together with an ice age that was already in the making, Toba may have accelerated cooling and drying worldwide, and driven global temperatures down as much as 27° F. This, in turn, dropped mountain snow and tree lines by nine thousand feet, plunged the planet into a six-to-ten-year volcanic winter and possibly an additional one thousand-year cooling episode.

As you might imagine this would have made life for the human species that were alive at the time even tougher than it already was, especially if they were living west and downwind of the eruption. The immediate effect would have been to drop uncounted cubic tons of choking volcanic ash on everything for thousands of miles around. Studies show that an ash layer a half foot thick draped all of south Asia, and quickly blanketed the Indian Ocean, and the Arabian and South China seas as well.

A layer of ash this thick would have decimated plant and small animal life on land and sea for years, catastrophically rattling the food chain and every creature that relied on it for survival throughout Asia and into Africa. Recent fossil and genetic studies suggest that the populations of gorillas, chimpanzees, orangutans, and even cheetahs and tigers dropped to near extinction levels.

Neanderthals in Europe and west Asia were apparently spared the direct effects of the volcanic fallout. *Homo erectus* living at the time in east Asia and (possibly Australia), and *Homo floresiensis*, the "hobbits" who lived close by seemed to have escaped because they were upwind of the debris. They may all, however, have suffered at the frigid hands of the explosion's longer-term effects.

The humans who seemed to have been hit hardest by the remarkable eruption were our ancestors, pockets of *Homo sapiens* scattered throughout Africa. As the debris spread, some scientists believe the eruption's cooling effects nearly wiped us out, a genetic coup de grâce that would have made this book, and you and I, entirely impossible.

It's unlikely that Toba by itself can explain the sudden whittling of our ancestors around this time in prehistory, but it certainly didn't help. Except in one surprising way. By isolating *Homo sapiens* settlements and placing even more survival pressure on them, it may have led to hardier and more adaptable men and women. There could be something to that hypothesis because while other primate species seem to have slowly rebounded, *Homo sapiens* not only bounced back, its population exploded and began to move quickly into Asia, Europe, and the remainder of the planet.

began to flow back into one another, setting the stage for a remarkable migration that changed the world.

Mitochondrial genetic studies tell us that around this time one small group of modern humans living in Ethiopia or the Sudan, armed with an assortment of high-tech tools—bone and ivory hand axes, long spears, and fire-hardened stone knives mostly—headed northeast, then over the Red Sea into Yemen on the Arabian Peninsula.

Undoubtedly, being human and curious, several waves of our direct ancestors ventured from their mother continent into the Mideast during this time. Modern human migration wasn't likely one solitary foray northward. During cold oscillations, seas everywhere would have grown shallower, including the Red Sea and the Gulf of Aden.

Right where these meet at a place with the dramatic name the Gate
of Grief (Bab el Mandeb), the continents of Asia and Africa nearly
kiss. Even today the distance between them is slender: no more than
twenty miles of seawater divides the huge continents. But during
frigid climatic swings the Red Sea sometimes dropped more than 210
feet, narrowing the straits by several more miles, nearly attaching Africa and Asia like two continent-size Siamese twins.

Though there is no evidence the seaway ever completely dried
out, climatologists believe that a chain of small islands sometimes
emerged between the immense landmasses. The shorelines of those
islands would have made excellent places to fish and eat before moving northeast to the next island. In time, the traveling tribes inevitably
made their way, perhaps in small boats or on rafts, island by island, to
the underbelly of Asia.

This feat, however long it ultimately took, liberated our kind
from the boundaries of Africa to head off to every continent and
landmass the planet had to offer. We weren't the first to find a way to
Asia, but this migration, or series of them, would, in the astonishingly
short space of fifty thousand years, utterly revamp an entire planet.

Once on the mainland of Asia, modern humans began to ripple
outward. Some bent east, hugging the shorelines of Yemen, Oman,
Iran, and Pakistan as they headed toward India, and others migrated
north through Mesopotamia (Iraq). There, this second group split
again, some working their way from Turkey through the Danube cor-
ridor, others sticking closer to the Mediterranean coast as they headed
toward Greece and the boot of Italy.

The splitting branches of humans grew like a bush. The fossil and
genetic evidence tells us that within five millennia our kind had settled the ancient continents of Sunda and Sahul, landmasses that exist
today as the oceanbound islands of Indonesia and New Guinea and
the continent of Australia. Forty-five thousand years ago, however, the
Indian Ocean was shallower, and these exposed shelves of land were
separated by straits no more than sixty miles at their widest. This means
that inside of ten thousand years, wandering *Homo sapiens* doggedly
walked from the eastern Sahara to the plateaus and mountains of western
Australia.

Meanwhile, other branches of our kind that had radiated into Mesopotamia and moved west spent the next fifteen thousand years settling much of Europe, as far away as Spain and well north of the Alps.

Paranthropus aethiopicus
This creature was among three species of "robust" humans, some of whom
roamed the plains of Africa for as many as a million years. They might
have outcompeted the line of primates that eventually led to us, but our
direct ancestors took an odd evolutionary turn that lengthened our childhoods and profoundly changed human evolution. (See Chapter 2: "The
Invention of Childhood.") *Original artwork by Sergio Pérez.*

Homotherium—Big Cat of the Ancient Savanna

Life on Africa's ancient savannas had to have been terrifying. The humans who roamed and foraged there between 5 million and 1.5 million years ago probably spent a good deal of their time avoiding big cats like this one, a precursor of today's lions, panthers, and tigers. The danger they presented further bonded early humans, making cooperation among them more important than ever, one reason we are so social today. (See Chapter 2: "The Invention of Childhood.") *Homotherium © 2005 Mark Hallet.*

Lake Turkana—An Evolutionary Garden of Eden?
Today Lake Turkana is the world's largest alkaline lake, but millions of years ago it was the garden spot of Africa and home to ancient human species of all kinds, including the line that likely led to us. (See Chapter 3: "Learning Machines.") *Photo credit: Yannick Garcin.*

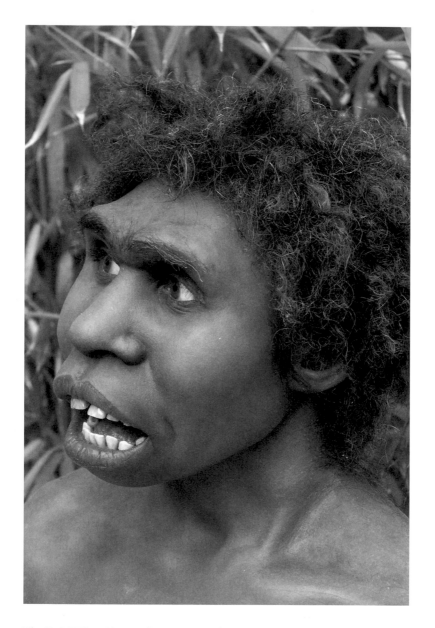

The Boy Who Changed Our View of Human Evolution
Also known as Nariokotome Boy, this young man met his end 1.5 million years ago. Luckily, and remarkably, most of his skeleton survived, making him one of the most important paleoanthropological finds ever. His teeth and bones have illuminated the mysterious evolution of our long childhoods and the crucial role it played in our survival. (See Chapter 2: "The Invention of Childhood.") *Photo credit: Look Sciences/Photo Researchers.*

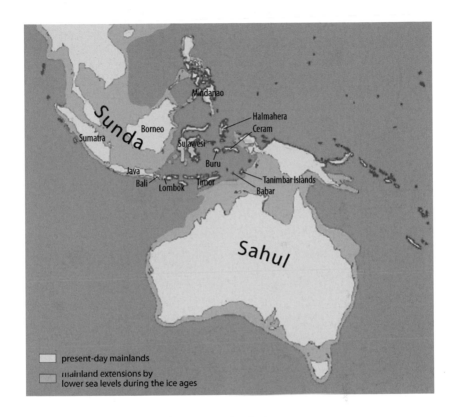

present-day mainlands
mainland extensions by
lower sea levels during the ice ages

The Ancient Continents of Sunda and Sahul

Fifty thousand years ago waves of modern humans began making their way out of Africa, scattering in all directions. Some tribes wandered to Australia, more than ten thousand miles away. They were able to make that journey because forty-five thousand years ago a swing to frigid climate locked oceans of water in earth's polar caps and dropped global sea levels. That created these immense continents in the Indian and Pacific oceans which are today submerged: Sunda and Sahul. Across these landmasses (with some short hops by sea) early humans made their way to the plateaus and mountains of western Australia, the ancestors of today's Australian Aborigines. (See Chapter 5: "The Everywhere Ape.") *Based on original artwork by Maximilian Dörrbecker.*

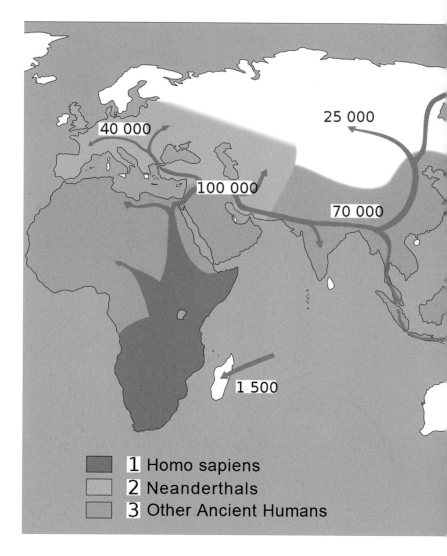

The Scattering of the Human Race
Once they had departed Africa, modern humans headed off to every corner
of the planet—the Middle East, Europe, Asia, the Far East, the South Pacific, Australia, and the Americas. Among the last continents to be reached?
Antarctica, in the nineteenth century. Remote Pacific islands were probably
populated about the time the first Pharaohs ruled Egypt. (See Chapter 5:
"The Everywhere Ape.") *Original artwork by Altaileopard, Wikimedia commons.*

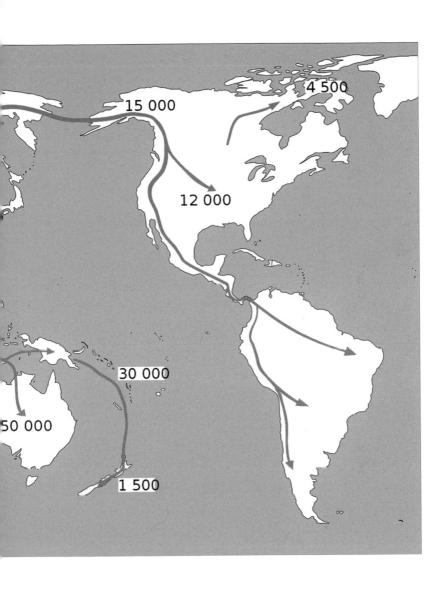

Gorham Cave
Twenty-five thousand years ago the last Neanderthals may have lived, and died, in this cathedral-like cave. (See Chapter 6: "Cousin Creatures.") *Original photo provided by Gibmetal77, Wikimedia commons.*

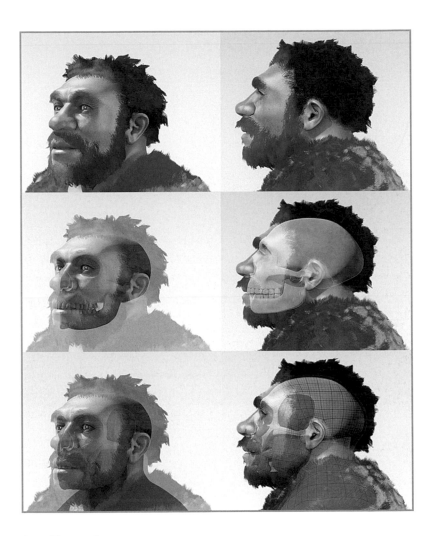

Our Closest Cousin?
We now know the Neanderthal people of Europe and west Asia were
remarkably intelligent and tough. This reconstruction illustrates that their
large skulls, thick, ropey muscles, and expansive noses, optimized for
warming cold air, helped them survive frigid temperatures and a punishing lifestyle. (See Chapter 6: "Cousin Creatures.") *Original artwork by Cicero
Moraes, Wikimedia commons.*

Final Days of the Neanderthal
Did the last Neanderthal sit on the great snaggled-toothed Rock of Gibraltar
and watch her (or his) final sunset? (See Chapter 6: "Cousin Creatures.")
Original photo provided by RedCoat, Wikimedia commons.

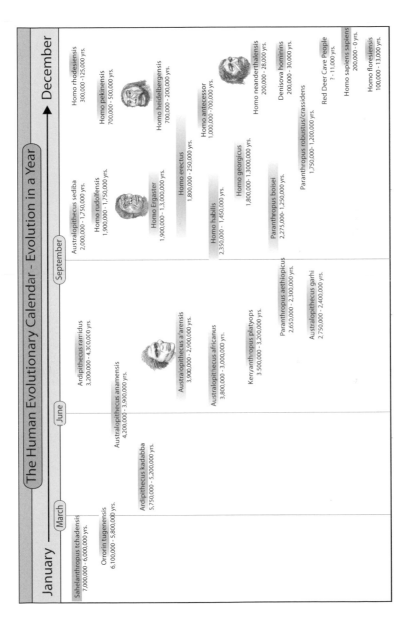

The Human Evolutionary Calendar - Evolution in a Year

January — March — June — September — December →

Sahelanthropus tchadensis
7,000,000 - 6,000,000 yrs.

Orrorin tugenensis
6,100,000 - 5,800,000 yrs.

Ardipithecus kadabba
5,750,000 - 5,200,000 yrs.

Ardipithecus ramidus
3,200,000 - 4,300,000 yrs.

Australopithecus anamensis
4,200,000 - 3,900,000 yrs.

Australopithecus afarensis
3,900,000 - 2,900,000 yrs.

Australopithecus africanus
3,800,000 - 3,000,000 yrs.

Kenyanthropus platyops
3,500,000 - 3,200,000 yrs.

Paranthropus aethiopicus
2,650,000 - 2,300,000 yrs.

Australopithecus garhi
2,750,000 - 2,400,000 yrs.

Australopithecus sediba
2,000,000 - 1,750,000 yrs.

Homo rudolfensis
1,900,000 - 1,750,000 yrs.

Homo Ergaster
1,900,000 - 1,3,000,000 yrs.

Homo erectus
1,800,000 - 250,000 yrs.

Homo habilis
2,350,000 - 1,450,000 yrs.

Homo georgicus
1,800,000 - 1,3000,000 yrs.

Paranthropus boisei
2,275,000 - 1,250,000 yrs.

Paranthropus robustus/crassidens
1,750,000 - 1,200,000 yrs.

Homo rhodesiensis
300,000 -125,000 yrs.

Homo pekinensis
700,000 - 500,000 yrs.

Homo heidelbergensis
700,000 - 200,000 yrs.

Homo antecessor
1,000,000 - 700,000 yrs.

Homo neanderthalensis
200,000 - 28,000 yrs.

Denisova hominins
200,000 - 30,000 yrs.

Red Deer Cave People
? - 11,000 yrs.

Homo sapiens sapiens
200,000 - 0 yrs.

Homo floresiensis
100,000 - 13,000 yrs.

If we could compress the emergence of all of the humans we so far know of who evolved over the past seven million years into the space of twelve months, it would look something like this. Many more species probably came and went that we haven't yet discovered. (See Chapter 1: "The Battle for Survival.") *Artwork and graph by Frank Harris, 2012.*

Prehistoric Genius

Long ago a Cro-Magnon artist painted this breathtaking image deep in the Altamira caves of Spain. Today they would be the envy of art galleries around the world, or Madison Avenue marketeers—rich, vibrant, and ingenious. You can almost see the image ripple in the ancient firelight that once illuminated it. Around this time in human history there was a global blossoming of creativity. Was the wellspring of that creativity our long childhood? (See Chapter 7: "Beauties in the Beast.") *Photo credit: akg-images.*

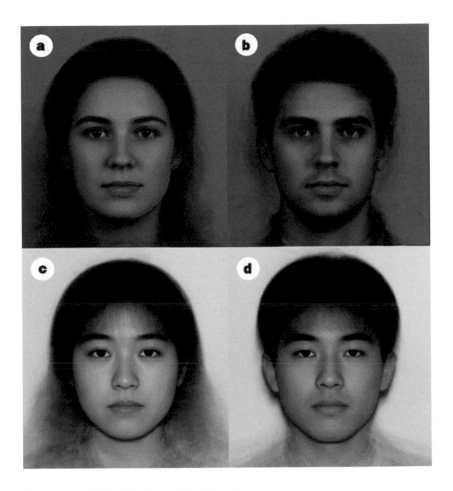

One Reason Why We Resemble Baby Apes

The effect of youthful (more feminine) faces on members of the opposite sex illustrate that even today both men and women find their counterparts more attractive if they look more childlike. For this experiment, scientists digitally created an "average," but attractive, version of two faces for each sex, one Caucasian and one Asian, four "average" faces in all. The researchers then digitally modified each face to create two versions, one slightly more masculine, the other slightly more feminine and childlike. (See Chapter 7: "Beauties in the Beast.") *Reprinted by permission, Macmillan Publishers Ltd: Nature 394, "Effects of Sexual Dimorphism on Facial Attractiveness," pp 884–87, August 27, 1998.*

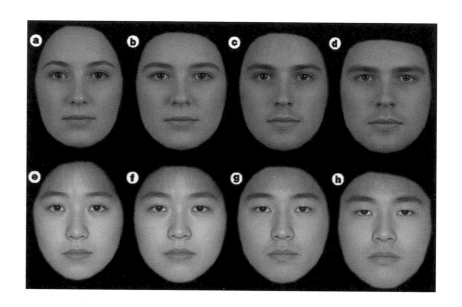

Our Preference for Childlike Looks Persists Today

The male versions in the study sport slightly heavier eyebrows, a hint of shaved beard, squarer jaws, and pupils that stand a bit farther apart than female pupils. This creates the illusion that the male faces are larger than the women's (they aren't). Ancient preferences like these help explain why we look, even in adulthood, more like baby apes than fully grown ones. (See Chapter 7: "Beauties in the Beast.") *Reprinted by permission, Macmillan Publishers Ltd.: Nature 394, "Effects of Sexual Dimorphism on Facial Attractiveness," pp 884–87, August 27, 1998.*

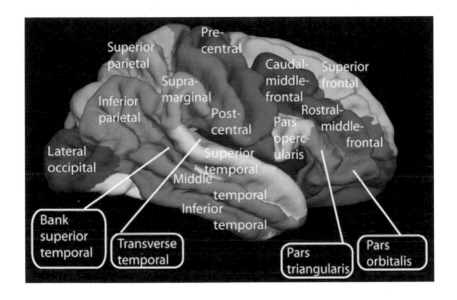

The Black Box We Call the Human Brain
The human brain is an amalgamation of ancient and newly evolved "mini brains," each with its own functions, cobbled together by the demands of evolution. Together they create the behavior we call human; complex, mysterious, playful, and unpredictable. Can the mind that the human brain makes possible comprehend itself? (See Chapter 8: "The Voice Inside Your Head.") *Original artwork by permission: Patric Hagmann et.al., Wikimedia commons.*

The Red Deer Cave People

Recently scientists stunned the world with the discovery of a mysterious people exhibiting both ancient and modern features who lived in southern China as recently as eleven thousand years ago, just as Homo sapiens were inventing agriculture. Are they somehow related to us, Neanderthals and the newly discovered Denisovan people, or are they an entirely separate branch of the human family tree? Recent discoveries have rapidly rearranged old assumptions. More changes will likely come. (See Chapter 6: "Cousin Creatures.") *Original artwork by Peter Schouten.*

Still others expanded north and east into Asia, across the highlands of Tibet and the steppes of Russia until they had nearly reached the top of the world to cross the land bridge between Russia and Alaska. From there they made for Canada, into North America and then southeast to the Meadowcroft settlement just outside Pittsburgh, Pennsylvania, sixteen thousand years ago, and finally into Central and South America to become, someday, the Incans, Mayans, Aztecs, Hopis, and scores of others. The settlers of Meadowcroft predated European explorers by a mere 155 centuries.

The upshot of all of this restless meandering was that after millions of years of evolution, in the tiny space of fifty millennia, the newest addition to the human family tree had wandered into and settled two Americas, all of Europe, Asia, Africa, Southeast Asia, even Japan. Only Micronesia and remote islands scattered across the Pacific, places such as Tahiti, the Philippines, and Hawaii, remained uninhabited. It took another two thousand years, as civilizations in Egypt, Mesopotamia, China, and India were rising, for small groups of dauntless explorers to travel across thousands of miles of open sea to populate, for reasons we have yet to fathom, those tiny specs of land.

Though the time scales we are talking about—mere tens of thousands of years, rather than hundreds of thousands or millions—seem relatively short, they still dwarf the length of recorded human history. Civilizations in India, China, and Egypt have come and gone within thousands of years. Alexander's empire disappeared within a few generations. Even Rome ruled most of Europe, parts of the Middle East, and North Africa for fewer than one thousand years. Ninety-five percent of the time that our particular brand of humanity has existed remains mysterious and almost entirely unrecorded. Nevertheless a great deal happened and a great deal of territory was covered.

Despite the mountains of research (and years of heated debate) that focus on the departure from Africa of our species, it's important to remember that we were not the first humans to forsake the continent for other parts of the planet. But before we get too deeply into the specific and considerable wandering of earlier humans, it might be best to clarify some recent, and surprising, rearrangements of our family tree.

For decades paleoanthropologists generally agreed that *Homo erectus* was a single species from which we evolved directly, and creatures we now know as *Homo ergaster* were assigned to the *erectus* line. But it now seems, at least by the lights of an increasing number of researchers, that *ergaster* represents a species of its own, and the primates we have been calling *erectus* are actually a potpourri of many human species whose bones simply haven't turned up in great enough numbers to warrant their own names or branch on the family tree. Now paleoanthropological sentiment now seems to be leaning toward our being descended, a few species removed, from *ergaster*, while *erectus*, and other humans who left Africa and wandered east, became one or several other species that eventually died out.

Whether or not we are descended directly from them, the species we collectively refer to as *Homo erectus* rose up and began to fan out to the mid and far east in small, tight brigades as long ago as 1.9 million years. Very adventurous to place some temporal perspective on this, consider that this is one million seven hundred thousand years before we *Homo sapiens* even showed our face. Or put another way, eight hundred times the twenty centuries that have passed since Augustus Caesar ruled the Roman Empire. *Homo erectus*, and the cousin species they left behind in Africa, were wickedly smart for their time and armed with the greatest evolutionary advancements of their day. Not bigger fangs, shaper claws, or stronger bodies, but a knack for thinking on their feet, working together, and adapting on the fly.

As a group they were tall, slim hipped, and built for long-distance running in the unforgiving sun of the equatorial lands where they arose. Their elongated limbs exposed the maximum amount of body surface to the air to help keep them cool. They were probably nearly hairless by this time, able to perspire much the way we do (chimpanzees have about half the sweat glands we do, mostly on their hands and feet, which aren't covered with hair), and outfitted with a complex network of blood vessels in their rather large heads, which helped efficiently vent heat and thereby avoid death by heatstroke. They were more adept toolmakers than *Homo habilis* before them and carried, by all accounts, an entirely new kind of technology with them everywhere, the same way today we tote around cell phones—the Acheulean hand ax, which was something like a Neolithic Swiss Army knife. Soon they would tame fire.

Before long, wave after wave of *Homo erectus* were settling parts of Arabia, China, India, even Indonesia, though they apparently never made it as far as Australia. Perhaps they weren't quite gifted enough to develop the advanced seafaring skills they needed to make it that far, or maybe the erratic climate prevented them. Or perhaps many tried, but none succeeded, or maybe they did succeed but their remains have yet to be found. Whatever the case, they wandered and meandered and migrated hither and yon with a far less sophisticated toolkit than *Homo sapiens* did when they exited Africa 1.7 million years later. But migrate they did.

One branch of the *Homo habilis* family (or perhaps an early version of *Homo erectus*) managed to travel all the way to Dmanisi in the Republic of Georgia 1.78 million years ago. Once again the weather helped. Between 1.9 and 1.7 million years ago Earth's climate was enjoying a respite between glacial periods, and these humans might have made their way up what would have then been a lush Nile River valley and then east across the narrow straits of Suez to the Arabian Peninsula before heading north of the Black Sea. Another branch apparently walked through a "green Sahara" filled with tall grass, brimming with wildlife, then made their way to current-day Algeria to settle at a site called Aïn Hanech.

Something more than simply human wandering was afoot here. The lands where these creatures were settling not only stretched from Indonesia to North Africa, but represented a widening variety of environments. They ran the gamut from marshes and streams to seacoasts and wooded mountainsides. The creatures were not only putting more distance between themselves and their home continent, but between themselves and the dictates of their genes. They were using their brains and creativity to adapt. While the other great apes had for millions of years been retreating to increasingly smaller forests, sticking to familiar environments where they were comfortable and for which they were genetically suited, these ancient humans, armed with their tools, clothing, and that magical thing called fire, were adapting their new environments to them, not the other way around.

By a million to 1.3 million years ago, entirely new species began advancing into the cold climates of Europe, trudging as far north as the British Isles, from (scientists suspect) northwest Africa across the Straits of Gibraltar. These were a species paleoanthropologists call *Homo antecessor*, a toolmaking cave dweller with a brain three quarters the size of ours and a more human and less simian face whose remains were first discovered at a railroad cut in Sierra de Atapureca, Spain. Another species descended from *antecessor* known as *Homo heidelbergensis* may also have made his way to this same area. More on him soon. But in either case each represented newly minted humans, most likely descended from *Homo ergaster*.

Interesting evolutionary events had been unfolding on the plains of Africa for the past several hundred thousand years, making the story of our own emergence both more interesting and more murky. While brigades of various hominin explorers were fanning off in every possible direction, the root species (including *ergaster* and *erectus*) were continuing to diversify back on the home continent as well. Around seven hundred thousand years ago, an altogether new, and crucially important, creature called *Homo heidelbergensis* stepped out of the mists of time.

The remarkable thing about *heidelbergensis*, so named because the first specimen was found near Heidelberg, Germany, is that it is the species from which both we and Neanderthals descended. That news has utterly rearranged the human family tree. Until recently it was thought that we could not possibly share anything much, especially our ancestry, with these tough, burly creatures who passed into extinction some twenty-five thousand years ago. We were, according to common wisdom, directly descended from *Homo erectus*. Yet, it turns out, if not for *heidelbergensis*, neither we nor Neanderthals would ever have walked the earth.

The creatures, the people, who later evolved into *Homo sapiens* and Neanderthals began to part ways, genetically speaking, from *heidelbergensis* almost as soon as *heidelbergensis* itself emerged. Some members of the species remained on Africa's horn (and are sometimes referred to as *Homo rhodesiensis*), but others, with a more extreme case of wanderlust, moved northwest across a new green Sahara to Gibraltar and then into Europe, following in the footsteps of *Homo antecessor*.

The archaeological evidence suggests that these nomads became the first humans to build shelters, probably of rock and wood, and hunted big game, like Irish elk, mammoths, and European lions, with long wooden spears. These inventions served them well in the colder climates they were dealing with throughout Europe, especially when glacial ice descended. They had large brains—1100 to 1400 cc, as large as ours—and by this time were easily the brightest of the planet's primates. Even more than *antecessor*, the shape of the outer and inner ears of these people indicates that they could make fine differentiations between sounds, a trait that leads some scientists to speculate they used sophisticated speech of some kind. Dental wear on teeth in the right side of their mouths means they may have been using their mouths as a "third hand" clenching tough food, tools, or clothing on their right as they worked on them. This could mean they were right-handed, and right-handedness is associated with language and lateralization of the brain.[5] Thin soup as scientific theories go, but something worth chewing on.

The original *heidelbergensis* was, it seems, thick boned, huskier and stronger than *erectus*, who was taller and slimmer. Not that at six feet he was short, but with a frame that easily supported two hundred or more pounds, he was built like a bouncer, or college fullback, a trait that mystifies scientists some-what since most African primates tended to be long of limb, the better to expose more of their body to the air, a form of natural air-conditioning.

Homo heidelbergensis

The European branch of *heidelbergensis* maintained and built upon these traits as they evolved into Neanderthals. The colder climate favored thicker, stockier creatures that exposed less of their bodies to the air. (Inuit people and the native residents of Siberia also show these same traits as ways to conserve heat.) The strength and endurance these bodies were apparently blessed with were certainly assets as they dealt with a punishing climate. Bulkier, stronger, physically tougher individuals would also have been favored when it came to hunting big game. As far as we know, these people and the Neanderthals that followed them did not throw their wooden spears when they hunted. Instead they used them to repeatedly jab their prey at close range, a dangerous way to shop for dinner.

You can imagine that this not only took courage and strength, but a body that could survive being tossed around by a wounded and en-raged lion, mammoth, or woolly rhinoceros and still bounce back.

Paleoanthropologists have found evidence all over the world of the beatings Neanderthals withstood. Skeletons found from the Middle East to Western Europe have revealed ugly injuries to their ribs, spine, lower and upper legs, and skull. What's more, these injuries usually healed and there are no signs of infection. More than once scientists have noted the injuries resemble the kinds of hammerings that rodeo riders sustain from big, bucking animals. Except in the case of Neanderthals they weren't riding bulls or horses, they were hopping on the backs of wolly rhinos, aurochs, or elk to jam their long spears in one killing blow through their back behind their necks. From time to time, of course, the animals they hunted might not have taken kindly to this.

Despite the beatings Neanderthals survived over the next half million years and spread throughout Europe, following the retreating glaciers north when temperatures moderated, and heading south when the glaciers returned. In time they became the dominant primates in Europe and settled it from the British Isles to the shores of the Black Sea.[6]

For the African branch of the family, life was challenging, too, but for entirely different reasons. Continuing increases in climate fluctuation meant surviving waves of crippling droughts. But their large brains, their tough bodies, and their increasingly strong social structure saw them through. In the end, both the Africa and European branches of humanity outlasted several climatic swings, until finally, around two hundred thousand years ago, they had completed their transformation into two entirely different, but enviably advanced, species—the first *Homo sapiens* and the first Neanderthals.

If you hold the fossilized skull of a Neanderthal in your hand and closely inspect it, you might find it difficult to believe we share a common ancestor, but time, climate, and random chance are powerful change agents. Their brow ridge is thick, their heads longer, shaped more like a watermelon than a cantaloupe, like ours. Their chins recessed, or more accurately the middle part of their face protruded more than yours and mine and looked more muzzlelike around the mouth and nose, which was large and fleshy and well rigged for warming the cold northern air they breathed. And they were stouter, bulkier, barrel-chested.

We were slimmer than they were, but not so much because *Homo sapiens* in Africa had grown more gracile over time; we simply didn't accentuate the robust traits we had inherited from *heidelbergensis* the way Neanderthals did. In fact, when the two species later met in Europe six hundred thousand years or so after the original *heidelbergensis* branches split, *Homo sapiens* were probably on average taller, if not stronger, than their Neanderthal cousins.

Climate made Neanderthals even huskier than their big-bodied ancestors. While their collarbones were long, their broad shoulders curved inward around a chest that was both broad and deep as if to better husband their body heat. Their fingers, which must have nearly always been exposed to the cold, grew stubbier and rounded at the tips, an antidote to frostbite. Their big upper bodies balanced on a pair of bowed thighs above Brobdingnagian knees and shortened shins. But this did not mean they walked hunched over, apelike. They didn't. Like us they stood fully upright and could walk and run just as well as we do. They were simply a human species optimized for the cold, remarkably strong and outrageously intelligent. And given their longevity, astute and wise in the unforgiving ways of survival.

By the time we and Neanderthals had emerged, at least four (and probably more) intelligent, self-aware human species were still living on planet Earth. (See sidebar "The Newest Members of the Human Family," page 90.) Each was colonizing settlements spread sparsely from Britain to Indonesia, and from the Balkans to the southern tip of Africa. We do know that *Homo antecessor* and *heidelbergensis*, and their precursors *ergaster* and *habilis,* had by now gone the way of the dinosaur, but *erectus,* or some version of it, still roamed Asia while *Homo sapiens* made its itinerant way around Africa, and Neanderthals ruled Europe and west Asia.

There were no census takers fifty thousand years ago, so we don't know how many humans were living on the planet, counting members of every species, though genetic studies may soon illuminate this; a few hundred thousand, perhaps, certainly less than a million. The generally accepted view is that we *Homo sapiens* bided our time in Africa until we launched a concerted worldwide migration beyond the Dark Continent beginning about this time. This is called, not surprisingly, the Out of Africa theory. According to this hypothesis *Homo sapiens* displaced and then eventually *re*placed all other human species that had arisen over the long epochs that preceded the post-African travels of *their* ancestors, whomever they might have been.

The Newest Members of the
Human Family

As I was writing this book, various teams of scientists around the world announced the discovery of four entirely new species of humans, an indication of exactly how quickly the field, and the human family tree that reflects it, is changing. (See The Human Family Tree, page 12.) Three of these were discovered the old-fashioned way—fossilized bones stubbornly excavated from their hiding places in the ground. Of those three, two lived some time ago—*Australopithecus sediba* and *Ardipithecus kadabba*—species that roamed Africa two and four and a half million years ago, respectively. From these remains paleoanthropologists have been able to develop some fairly deep insights into these creatures' anatomies and lifestyles.

Based on four partial skeletons found in South Africa, *sediba* illustrated an emerging theme in paleoanthropology: there was a good deal more variation in ancestral humans than previously thought, and therefore lots of room to debate where they fall in the family tree. *Sediba* seems to have combined some old australopithecine traits and some traits of early *Homo* species. Its brain wasn't terribly large (about 450 cc), but hand, pelvis, and leg bones indicate it may have been an early tool user and well on the way to walking upright more often than not. Yet fossilized plants found with some specimens tell scientists that sediba lived in forested areas as well as open ones and often ate fruits like their chimpanzee cousins.

Researchers read these clues in different ways. Some argue that *sediba* was a precursor to the *Homo* species of humans that followed. Others didn't believe this was possible because that branch had already sprouted on the human family tree a half million years earlier with the emergence of *Homo rudolfensis*.

Ardipithecus kadabba is an ancestral human, and lived as many as three and a half million years before *sediba* in Ethiopia. (Some debate this age and set it one million seven hundred thousand years earlier.) He is ancient enough that his big toe was still designed for grasping tree branches, though other aspects of his anatomy indicate he moved on two feet on open ground. His brain was about the size of a modern bonobo at 300 to 350 cc, but his smaller incisors indicate, at least to some paleoanthropologists, that he and his fellow creatures were more socially cooperative than chimps. Male chimps have large incisors often used when battling for the attentions of the troop's females. This makes sense if he was spending more time in the more dangerous open grasslands where he would have to rely more on others in the troop for survival.

The third and fourth species come from a different time and different parts of the world than the first two. Both dramatically reinforce the emerging reality that our direct ancestors coexisted with a variety of other extremely sophisticated humans throughout the world until very recently. Just a few years ago an assertion of this kind would have been considered heresy in the field of human evolution. Each of these species lived when we *Homo sapiens* did, and DNA evidence indicates that at least one also mated with us, and with Neanderthals. Human species embraced one another, it seems, in more than a metaphorical way when they had the chance.

Of these two the most recent discovery was announced in March of 2012, and because of its novelty remains controversial. The fossils haven't even yet acquired a scientific classification. Instead researchers call their find the Red Deer Cave people, humans, but not likely *Homo sapiens,* that lived in south central China, north of Vietnam, as recently as 11,500 years ago. That these people were setting up camp not long before *Homo sapiens* had made their shattering transition from hunting and gathering to agriculture is one of the aspects of this discovery that has anthropologists both giddy and astounded.

The astonishment, however, only begins there. The fossils reveal that these people looked a little bit like us but also like more ancient humans. They have our rounded brain cases, less sloped than Neanderthals, but still retained thick, simian-style brow ridges. Like ours their faces were flat and tucked under their brain, but their chin, though the jaw juts forward, isn't squared off like ours. And strangest of all, scans of their brain cases indicate that they had modern frontal lobes, but archaic parietal lobes, which sit farther back in our brain. It makes one wonder if their reality was different from ours, and if it was, how?

So where did these remarkable people come from? Scientists have speculated along three lines: They might have been descended from a group of *Homo sapiens* that departed Africa earlier than generally thought and survived and evolved in isolation. They may truly be an entirely different human species, like Neanderthals, people who evolved from an earlier branch of the human family tree, *Homo heidelbergensis* or *Homo erectus*, perhaps. Or they could be hybrids: *Homo sapiens* that mated with archaic humans who were also living in south China, something that might help explain their unusual mix of features.

The fourth and perhaps the most intriguing species recently discovered left behind almost no evidence of its existence; no clues about how it looked, what tools it used, or where it came from; hardly even a bone. Like the Red Deer Cave people it has also not yet been assigned a scientific name. Instead researchers refer to this species as the Denisovans because the two tiny fossils they *did* leave behind—a wisdom tooth and the tip of a pinkie finger—were found in Denisova Cave in the remote Altai Mountains of Siberia. You could hardly imagine more meager leavings. Yet, after scanning the mitochondrial DNA within these tiny specimens, scientists at the Max Planck Institute for Evolutionary Anthropology managed to decode the creature's entire genome. And when they had they realized that the juvenile to whom these paltry fossils had once belonged represented an entirely new human species that had hunted and settled in these mountains forty thousand years ago. Amazingly, Neanderthals, *Homo sapiens*, and Denisovans each lived in the very same cave, though probably not at the same time. The DNA analysis also revealed that the peoples who became *Homo sapiens*, Neanderthals, and Denisovans all shared a common ancestor a million years earlier. It's not yet known exactly which species that was, possibly *Homo ergaster*.

It turns out that we share a genetic link with Denisovans in another remarkable way. In analyzing Denisovan DNA the scientific team compared it with living humans from six groups: the !Kung people of South Africa, Nigerians, the French, Papua New Guineans, Pacific Bougainville Islanders, and the Han Chinese. They were electrified when they found that between 4 percent and 6 percent of the genomes of the people of Papua New Guinea and Bougainville Island contain Denisovan DNA. Scientists surmise the genes were introduced to the islands when the hybrid descendants of *Homo sapiens* and Denisovans migrated into Southeast Asia and later Melanesia. There is even some evidence that these descendants made their way to Australia and the Philippines.

It's difficult to not be transfixed by these discoveries when you really take the time to think about it. Like Neanderthals and *Homo floresiensis* they are species who fought and struggled and lived sophisticated lives for tens, even hundreds of thousands of years alongside our direct ancestors on the same planet we inhabit today. And if that isn't astounding enough, some even mated with our kind, contributing forever to our DNA. Were these aberrations or the norm? How many more species and hybrids might we find now that DNA analysis has opened so many genetic doors?

If that's true (and there's little debate that it is, though it's becoming clear it wasn't quite this simple), most of the different varieties of humans, given their nomadic ways, must have crossed paths from time to time as they wandered into the edges of one another's territories.

There is evidence of this in the rocky hills of Galilee, not far from Nazareth, the birthplace of Jesus Christ. In 1929, in caves that pock the hills of Qafzeh, Israel, two scientists found an ancient burial ground and, remarkably, the bodies of eleven anatomically modern humans. At first scientists thought the bones were no more than fifty thousand years old, but later, improved dating technology revealed that they were nearly twice that age, making them the oldest modern human fossils to be found outside Africa. As researchers continued rummaging through the site, they realized the bodies retained some of the more archaic features of their ancestors, but that they were culturally advanced. The ornamental shells and red, yellow, and black ocher paints they left behind indicated as much. So did the hearth and the burials of the bodies themselves, one of which included a mother and her child. Their tools, however, weren't as advanced as later *Homo sapiens'*.

The odd thing about that was that their tools instead resembled Neanderthal implements, yet they themselves were not Neanderthal. The best guess is that somehow they, or earlier generations, had crossed paths with their northern cousins and borrowed some of their technology because it was better than their own.

For all we know, these were early *Homo sapiens* explorers, the Marco Polos and Vasco da Gamas of their day, wandering the Arabian Peninsula while less adventurous *Homo sapiens* tribes remained on the home continent. By all accounts, their expedition wasn't terribly successful.

The skeletons of red and fallow deer, small animals, aurochs, and some seafood shells show they gave colonizing the area a game effort, but their excursions never made it beyond the hills of Qafzeh and barely beyond the borderlands of Africa. There is no evidence that any of their kind ever made it north of this sector of the Middle East, not this far back in human prehistory. Maybe the explorers retreated home across the Red Sea straits, played out and tired; or maybe those eleven buried were put to rest by a last few survivors, or maybe they hung on for years in a small group like a prehistoric version of the settlers at Plymouth Rock or Jamestown, until disaster or disease at last carried them off. No one knows.

In the same area, paleoanthropologists recently discovered that Neanderthal explorers likewise found their way south into Galilee thirty thousand years later (December 26 or so in the HEC), but they came from the north rather than the south. Did the fair-haired, bulky colonists run into lithe, dark-skinned people from across the Arabian straits? If so, did the Neanderthals do them in or run them off the peninsula back to Africa?

"At that point," says paleoanthropologist Nicholas J. Conard of the University of Tübingen in Germany, "the two species are on pretty equal footing." The tools of both *Homo sapiens* and Neanderthals would have been about equally advanced, and given what the Neanderthals had been facing in the wilds and weather of Europe the past 130,000 years, they would have been an extremely tough breed. Modern humans may not have been their match, not yet. Or perhaps they mated and their offspring dissolved into the continent and disappeared from the map. Either way, it seems that for another twenty thousand years or so *Homo sapiens* ceded Asia to their barrel-chested cousins.

Whatever happened, we do know that Neanderthals and *Homo sapiens* eventually encountered one another in Europe sometime after our long-lost ancestors finally made their big push out of the Dark Continent. But what about the East and the *erectus* bands that had begun heading off toward India and China and Southeast Asia two million years earlier? What became of them and their ancestors, and did ours make contact with them?

Scientists haven't uncovered direct fossil evidence of even a single meeting—no burial sites, artifacts, or bones—but in 2004 a team that included biologist Dale Clayton and anthropologist Alan Rogers, both working at the University of Utah, proved that our ancestors did indisputably have a close encounter with another human species in the Far East sometime around twenty-five thousand years ago. How could they know if there was no fossil evidence?

Head lice.

Like every other living thing on Earth, head lice have DNA. And like humans or finches or predatory big cats, different species of lice have different DNA. Anytime we find head lice on ourselves— breakouts among schoolchildren are more common than parents would prefer—we find two kinds that are rarely separated. Despite nearly always being in one another's company, however, each initially evolved separately while dining on two different species of early humans. One of those species led to us. The other is extinct. For those two species of lice to coexist today, both had to have come into close contact sometime in the past.

By studying their DNA and then time-stamping the evolution of both strains, the Utah study concluded that at least one meeting took place sometime between thirty thousand and twenty-five thousand years ago in Asia. "We've discovered the 'smoking louse,'" Clayton wryly observed. "The record of our past is written in our parasites," added Rogers.

What makes this discovery especially surprising, aside from its creative use of parasites to track human behavior, is that most paleoanthropologists believe that *Homo erectus* met his end seventy thousand years ago, long before this encounter could possibly have taken place. Nevertheless it's difficult to dispute the evidence. Parasites reflect the evolution of their hosts. They rely on them for their livelihood after all, and their fortunes and survival are inextricably bound. So some direct descendant of *Homo erectus* must have survived forty-five thousand years longer than previously believed. Whoever this species was, the genetic history of the head lice that colonized it shows that it split into two species around 1.18 million years ago, about the same time that *Homo erectus* and our direct ancestors in Africa, possibly *Homo ergaster,* parted ways. That explains why the lice themselves also parted company and eventually evolved into two species in the first place.

The lice reveal something else fascinating (who knew the little buggers could be so informative?). The *Homo sapiens* strain corroborates evidence that our direct ancestors had been reduced to extremely small numbers between one hundred thousand and fifty thousand years ago before rebounding and rapidly expanding, with their head lice, to colonize the rest of the world. This supports mitochondrial genetic evidence that our kind nearly met an early and tragic (at least for us) end around seventy thousand years ago before recovering to spend the next fifty thousand years becoming the planets dominant species on Earth.

Strangely enough, the archaic lice, the ones that made their homes on the heads of the species no longer with us, show not an iota of evidence that they went through either a similar bottleneck or population explosion. During those years, when *Homo sapiens* had nearly been rubbed out, perhaps by the Olympian-scale eruption at Lake Toba in Indonesia, these other humans were apparently getting along just fine. One theory is that they were safely upwind of the explosion and didn't feel the immediate, violent effects, though this doesn't explain how they managed to survive the subsequent global volcanic winter some scientists feel resulted from the gargantuan blast. All indications are that this line of humanity did just fine, at least until they crossed paths again with the *Homo sapiens* descendants of the species that they had split off from more than a million years earlier.

It is not as crazy as it might once have been thought that a more modern descendant of *Homo erectus* was still alive as recently as twenty-five thousand years ago. The more scientists examine the past, the more surprises they find. They found a particularly big one when the remnants of an entirely new human species that no one had had the slightest inkling had ever existed came to light in 2004 at Liang Bua, a cave on the island of Flores, 388 miles east of Java in Indonesia. After much debate and head scratching, most paleoanthropologists agreed that *Homo floresiensis*, as these remarkable creatures came to be known, was a bright, toolmaking human. The big surprise, beyond the discovery that these people existed at all, was their startlingly Lilliputan stature. The press and even astounded scientists took to calling them "hobbits." One three-foot-three-inch-tall adult-woman skeleton that was discovered turned out to be even shorter than Lucy.

Their brain size at 420 cc was also not much larger than Lucy's, a hominin that had walked the earth more than 3 million years earlier. Yet these creatures could control fire, make sophisticated tools, and hunt game, though it's still an open question as to whether they could speak or used any advanced language. How, scientists have wondered, could a species with a brain less than one third the size of ours pull off these sorts of impressive feats?

Our best evidence indicates that the Flores hobbits lived between ninety-five thousand and seventeen thousand years ago, the descendants of earlier *Homo erectus* settlers who were eventually reduced in size by an odd evolutionary phenomenon scientists call island dwarfing. Island dwarfing happens when natural forces cause species to shrink in size over time in isolated locations, presumably because resources are severely limited. The theory is that in a kind of ecological bargain, animals grow smaller rather than starve. By reducing their size, both resources and diversity are both preserved, and life goes on with predator, prey, and the entire ecological niche surviving in a sort of pygmy state. Dwarfing can have other advantages under these circumstances. It's easier to stay warm or cool when you are smaller, which saves energy and requires less food. On Flores, in addition to the hobbits themselves, scientists have found examples of a small, elephant-like creature called *Stegodon*, an animal the hobbits apparently hunted with some enthusiasm.

Because of *Homo floresiensis'* size, especially the size of its brain, scientists have enjoyed some spirited debate about whether it came to the island in the form of a lean and tall *Homo erectus* (remains of *erectus* have been found on nearby Java), then shrank over time due to island dwarfing, or whether it may have been the descendant of smaller, Lucy-size creatures who came out of Africa before *erectus* and then made their way somehow to the islands of Indonesia.

Could a smaller, less intelligent species such as *Homo habilis* or *Australopithecus afarensis* have made the ten-thousand-mile journey by land to Flores without the benefit of fairly advanced tools? It would be a remarkable feat. Their brains were considerably smaller and considerably less sophisticated than all varieties of *Homo erectus*. It seems a stretch that such wanderers would have evolved to develop the sort of technology scientists found on the island without the benefit of their brains' growing larger and more complex beforehand. It's more likely that somehow their brains had advanced to the sophisticated wiring of *Homo erectus*, at least, and then grown mysteriously smaller while not giving up the advantages of that wiring. In other words the brain grew tinier, but its complex architecture remained intact, like the perfectly replicated miniatures of homes and furniture you might see in a history museum.

The current consensus is that the last hobbit departed about seventeen thousand years ago, but some have speculated they may have lived on. Anthropologist Gregory Forth has hypothesized that Flores hobbits might be the source of stories among local tribes about the Ebu Gogo, small, hairy cave dwellers who supposedly spoke a strange language and were reportedly seen by Portuguese explorers who came to the islands in the early 1600s. Henry Gee, a senior editor at *Nature* magazine, has even opined that species like *Homo floresiensis* might still exist in the unexplored tropical forests of Indonesia.[7]

It makes you wonder how many other human species we may find as we comb through the planet. Could small pockets of *erectus* descendants have managed to survive in remote areas throughout Asia, or even made their way to North America? Could there be something to the sightings of yeti in the Himalayas or Big Foot in the American West after all?

The point is that almost anything is proving to be possible when it comes to human evolution, even hobbits, and if they nearly survived until the first great agricultural civilizations began to gain a toehold, then could the descendants of *Homo erectus*, whatever we might call them, have remained abroad for our ancestors to meet as they trekked through Asia on their way to Indonesia and Australia?

Possibly a larger, more evolved version of *Homo floresiensis* had survived Toba and the ice ages that battered the Neanderthals in Europe and reduced *Homo sapiens* to a few clans hanging on by a wispy thread in a drought-ridden Africa. It would have been no mean feat to survive that ice age, but maybe in Southeast Asia, on the ancient continent of Sundaland, life was less deadly than in other parts of the world. It could even be that the species from which we acquired the second brand of head lice we carry around with us today are a gift from the hobbits themselves, Denisovans, or the newly discovered Red Deer Cave people of China.[8]

For now we can only speculate, but that we met these people—whoever they were—and that they so generously shared their parasites with us indicates that our encounter was of the close kind. Tight quarters are generally required when divvying up lice. Unfortunately, there is no way to decipher exactly what variety the close encounters were. Possibly we killed the people and took their clothing, and the bugs came in the bargain. Murder on a large scale *has*, unfortunately, been known to take place when a new, powerful group of humans finds less technologically advanced people. We don't have to look any further than the wrecked civilizations of the Incas and Mayans in South America, Aborigines in Australia, and Native Americans in the United States for proof. It is also possible we simply colonized the same space and outcompeted them for limited resources with better hunting strategies, better tools and weapons, and more elaborate cooperation. Or maybe we mated with them, either forcibly or affectionately, or both. We may even have run across them when they had reached the end of their evolutionary rope, and their parting gifts to humanity were a bloodthirsty bug and a few hunting grounds.

Probably, whoever they were, they were not as cerebrally gifted as the *Homo sapiens* they crossed paths with. But that doesn't mean they weren't bright. They were certainly far more intelligent than today's chimpanzee or gorilla, which are devilishly clever in their own right. If they were directly descended from *Homo erectus*, they may have lacked advanced language. *Homo erectus* is unlikely to have mastered the spoken word, though he may have used complex gestures or other vocalizations to communicate. Speech and language are not always the same thing, as the thousands who speak American Sign Language can attest.

It's difficult to imagine how we could ever decipher how these people communicated. The business of unlocking the past without the benefit of a working time machine makes science uncertain, especially when it deals with the spoken word. Any encounter between our kind and this other branch of the human family, each of whom had been traveling quite different evolutionary roads for nearly two million years, must have boggled both of their minds when it finally took place.

You might compare the meetings to those between the civilizations of the Old and New Worlds five hundred years ago, even if the comparison isn't altogether accurate. Francisco Pizarro's clashes with South America's Incas, or the Iroquois's crossing paths with early French traders who were exploring northeastern America, or Captain James Cook's legendary encounters with the people of Polynesia, all brought together cultures that were radically different and fraught with misunderstanding, often tragic (Cook eventually met his end when he was hacked to death by the natives of Hawaii, who had come to realize he and his men were not the gods they originally thought they were). But at least the meetings were between two groups that were the same species! Their cultural experience was different, but their intelligence was the same. They both used language, they had each developed tools, and they had the same brains, genetics, and anatomies.

Nor, on the other hand, would the meetings have been anything like early recorded human encounters with Africa's apes. There would have been no mistaking even a friendly chimp for a member of the human race, never mind that we share nearly 99 percent of our DNA.*

When our direct ancestors came face-to-face with these other humans twenty-five thousand years ago, would they have seen them as equals, as an enemy, as nothing more than an interesting, or terrifying, animal? Would their cultures have been even remotely the same after two million years of genetic divergence? *Homo erectus*, we know, had tamed fire, like *Homo sapiens*, but their way of communicating must have been radically different. More different than that of a British naval captain and a Hawaiian chief. Had they developed music or art? Surely they were social. *Homo erectus* had, after all, evolved from the same gregarious stock we had, but how well organized were they, how complex was their society? Did they festoon or paint themselves? How did they dress? Had they developed religion or superstition to explain the world? Did they even care to explain it? Was there something about the chemistry or structure of their brains that made their reality fundamentally different from ours?

It's not a given that our kind would have dominated this other species when they did meet. A chimpanzee, despite its diminutive size, is strong enough that it can, rather literally, tear one of us limb from limb, if it chooses. And these people may not have been diminutive. Based on earlier fossils, it is entirely possible that they were faster, bigger, and stronger. *Homo erectus* men could easily reach heights in excess of six feet and could likely outrun our kind. (The same may have been true of the Red Deer Cave people, though we don't yet know enough.) *Homo erectus* was a species that had been around in one form

---

*When the ancient Carthaginian explorer Hanno the Navigator came across a group of what he called savage men and hairy women in West Africa twenty-five hundred years ago, he wasn't sure if they were human, but the difference between them and him was clearly large. His interpreters called the creatures *Gorillae*, from which we later derived the term *gorilla*. It's possible that's exactly what Hanno had encountered.

or another for nearly two million years, the longest run any human species has ever enjoyed based on the current, if sparse, information we have. The world had tested them again and again, and they had passed the test. When these people first spied the strange, globe-headed, square-jawed creatures with their throwing spears and fire-hardened tools, it must have been as shocking to them as having aliens from Tralfamadore beam down from the sky and show up in Times Square, a race of aliens with superior technology who had come seemingly out of nowhere. How would these people have explained one another to themselves?

It's fascinating to speculate on all of this, but, unfortunately, speculate is all we can do because, so far, like a crime without a clue, there is no archaeological evidence of the meetings. There are only the parasites. But what a shattering event that meeting must have been.

Our encounters with the ape-men of south Asia were not, however, unique. Twenty-five thousand years earlier, and half a world away, we came face-to-face with another branch of the human family tree, the native Neanderthals of Europe and west Asia. This time we were more closely related, and of similar intelligence. Here, thankfully, we have a little more hard evidence that can shed a bit of additional light on the nature of their astonishing encounters.

CHAPTER SIX

COUSIN CREATURES

*Ne·an·der·thal (nē-ā n'der-thôl', -tôl', nā -än' dēr-täl') also Ne·an·der·tal
(-tôl', -täl') — Someone who is big and stupid and thinks physical strength is
more important than culture or intelligence.*
—Macmillan Dictionary

MAYBE IT'S BECAUSE they aren't around any longer to defend themselves, but Neanderthals are among the most maligned species paleoanthropologists have ever taken to studying, and they have been studied since before there was any such thing as a paleoanthropologist. The first Neanderthal fossils to attract serious attention were found in 1856, a full three years before a nervous Charles Darwin had finally gotten around to sharing his provocative theories about natural selection with the publication of *On the Origin of Species*. The unearthing of the skull, torso, and legs that limestone workers near Düsseldorf in western Germany had shoveled onto a hillside made their long-deceased owner the first acknowledged representative of a prehistoric human species ever. Rather a big deal. Not that anyone realized this when the quarry's owner first examined the bones. Like the workers, he assumed these were the remains of a cave bear. Others speculated that they were what was left of a Mongolian Cossack who had failed to keep up with his fellow soldiers a few decades earlier when Russians were desperately fighting off Napoléon's army.

Luckily, rather than being tossed aside, and into oblivion, the fossils found their way to a local schoolteacher named Johann Carl Fuhlrott, who recognized immediately that they were human and got them into the hands of Hermann Schaaffhausen, an eminent anatomist of the day. After nearly a year's careful study, Schaaffhausen presented the bones to the rest of the scientific world and pronounced that they belonged to a savage member of a "very ancient human race."

Not everyone agreed. This was, after all, a time when many Europeans still held fast to the conclusion the Church of Ireland's Archbishop James Ussher had come to in 1650. God, he said, had completed the world's creation at precisely twelve o'clock P.M., October 23, 4004 B.C. The undisputed expert on human anatomy at the time, Rudolf Virchow, reckoned that because of the skeleton's unusual shape and the heavily ridged brow, these were the bones of a rickets-ridden, cave-dwelling hermit who had met an untimely death at the site sometime in the past, but not the deep, dark past.

That might have been the end of the whole discussion, but then in 1863, the highly respected British biologist Thomas Henry Huxley (of the remarkable Huxley family, which also produced Leonard, Aldous, and Andrew Huxley, among others) published his landmark book, *Evidence as to Man's Place in Nature*. Huxley was a devoted adherent of Darwin's theories, so devoted that in some circles he was known as Darwin's Bulldog. Being a bulldog, he made the argument that Neanderthals preceded modern humans somewhere down the line in our inexorable march from ape-like ancestors to our present form. In other words, he was an earlier version of you and me.

Homo neanderthalensis

Ultimately Darwin's and Huxley's views won out, at least generally and at least in the scientific world. Then in 1908, decades after the original discovery was made in Germany, France's leading biological anthropologist, Marcellin Boule, saddled the world with a damagingly inaccurate view of Neanderthal when he got hold of another set of bones that had been found in a rock shelter in La

Chapelle-aux-Saints in southwestern France. Boule studiously scruti-
nized the remains, but missed that the person to whom they belonged
had suffered from chronic arthritis and a disease that had cruelly twisted
the man's spine. So when he rebuilt the crippled Neanderthal's anatomy, the image he created was of an apish, bowed, and stoop-shouldered
creature who became the prototypical caricature of the caveman that
most of us still carry around in our minds—dim-witted, brutish, and
slow, something along the lines of a Harry Potter troll. His conclusion:
Neanderthals were not our ancestors but an evolutionary dead end,
which, oddly enough, turned out to be about right, but for all the
wrong reasons.

Insights into Neanderthals and their world have altered considerably within the past decade as new fossils have been discovered, and
scientists have applied genetic technology in creative ways to plumb
exactly who these remarkable people were. It's now clear that though
they lived under brutal and stupefying circumstances during their
nearly two hundred thousand years in Europe and Asia, they were
themselves neither brutal nor stupid. In fact their brains were slightly
larger than ours are today, and their accomplishments, when placed in
the context of the challenges they faced in their daily lives, were
nothing short of astonishing.

Two hundred millenia is a long time, and Neanderthals were by all
accounts a busy species throughout. Although the best evidence is
that their worldwide population never reached into six figures, they
still managed to range thousands of miles in all directions. The bones
of over four hundred Neanderthals have been unearthed during the
past hundred years. They reveal that at one time or another these
people lived as far west as the Iberian Peninsula and as far east as the
Altai Mountains in southern Siberia.[1] When the weather grew colder,
they traveled south to the Arabian Peninsula and Gibraltar, and when
glaciers receded, they receded with them up to the mountain ranges
of northern Europe. There is no evidence that they ever ventured
into Africa, which makes sense. Their bodies were optimized for cold
weather, and over the past two hundred millennia there was plenty of
that in Europe and their haunts in Asia.

Neanderthals' physical adaptations to the cold are among the reasons we think of them as brutish. On thick necks they carried large
heads to hold their big brains (one fossil cranium indicates a brain of
over 1700 cc, about 300 cc larger than your brain or mine). Their jaws

were big with long rows of square teeth, but their chins were small, almost as though the middle part of their face had been pulled out slightly around their nose and upper lip. (From their point of view, it would have looked as though ours had been pushed in and flattened.) The thick brow ridge that ran over their eyes gave them a brooding, almost sinister look, even if their heads were topped, as some scientists have speculated, with mounds of red or blond hair. Their hair color and their fairer, possibly freckled skin were an evolutionary accommodation to living farther north than the *Homo sapiens* from the warmer climates of the south. Dark skin in equatorial environments evolved to reduce the amount of vitamin D we absorb, but light skin increases our absorption rate, a good thing in lands where sunshine is in short supply for half the year.[2]

The selective pressures of cold, northern climates also endowed Neanderthals with big, rounded shoulders and thick-barreled chests that would shame a professional fullback. Even their noses helped them survive frigid temperatures. They were enormous and fleshy and rigged with expanded nasal membranes that warmed and moistened the cold, dry air they breathed. Above all they were strong, much stronger than we are today, with slightly foreshortened arms and thighs that reduced the amount of skin they exposed to the air. Their hands were large and far more powerful than their *Homo sapiens* cousins', and their forearms were thick and roped with muscle, at least if the anatomy of the fossil bones that ran from their wrists to their elbows are any indication.

Despite their rounded shoulders and foreshortened legs, the fossils scientists currently have in hand indicate they were not shorter than the *Homo sapiens* of their time, though they were shorter than their direct ancestor, *Homo heidelbergensis*, who stood six feet tall, and the slender *Homo erectus*, creatures who were as well optimized for running and hot climates as Neanderthals were for battling big game and cold weather. What they lacked in height they made up for in bulk, which may, in an odd way, have contributed to their undoing. To stay warm and maintain their enormous strength, some scientists have theorized, they required up to 350 calories more a day than their *Homo sapiens* counterparts. Today 350 calories might not seem like much, nothing more than an extra muffin at Starbucks, but fifty thousand years ago that much extra food would have been exceedingly difficult to come by day in and day out.

It's tough to find more persevering creatures than Neanderthals.

They survived the most punishing climate Europe could dish out for a length of time that dwarfs all of the history we have so far recorded hundreds of times over. They were clever, fierce, and successful hunters who could bring down deer, bear, bison, and mammoths. One site that dates back 125,000 years reveals that a group of Neanderthals living in a cave at La Cotte de Saint Brelade drove mammoths and rhinoceroses over a nearby cliff, butchered the dead or writhing animals on the spot, and then hauled in the choicest cuts into their nearby caves before any hungry predators could get to them. Efforts like that took brains and cooperation and sophisticated communication. Their culture was advanced and their social structure tight and fair, otherwise they would never have survived as long as they did.

The evidence from Shandihar Cave in Iraq indicates they began to bury their dead before we *Homo sapiens* did, going as far back as one hundred thousand years.[3] Long ago in a ceremony we can only imagine, fellow Neanderthals gently laid the body of a man to rest in a shallow grave, positioned fetal-like, as though he were sleeping. He had had a rough life. Multiple broken bones, degenerative joint disease, a withered arm, and an eye that was probably blinded all attest to that. Yet the pollen and the ancient remnants of evergreen bows that investigators found lying below and around him indicate that this man was loved and important to those who saw him off to death or, in their minds perhaps, to a new kind of life.

The same arthritic man that Marcellin Boule had maligned in 1908 as stooped and apish had also clearly been cared for by his fellow tribesmen. He was not young when he met his end, but forty to fifty years old, ancient by Neanderthal standards. Walking must have been agonizing given the state of his bones. He died with no more than two teeth, which would have made eating the normal, rough Neanderthal diet nearly impossible. Yet this man's fellow tribesmen must have carried and fed him specialized foods for years, otherwise he would never have lived to such a ripe age.

This gives us a peek into what the Neanderthal mind may have been like, but only a peek. Behaviors like these tell us that Neanderthals probably felt the loss of death, mourned those close to them who had met their end, and, by extension, understood there was something more to life than the day-to-day problems it presented. They, like our ancestors, must have wondered what follows death.

It's strange to think that a creature we have always seen as a club-wielding brute was more softhearted than we are. Again and again Neanderthal fossils reveal that these people took immense punishment—yet their wounds often healed, which means that their comrades did not leave them behind even if they were severely injured, but instead kept them in the clan and nursed them back to health.

It's not surprising that they were injured. The long hunting spears Neanderthals routinely used weren't the sort that could be thrown from a distance. (Most anthropologists hold that the Cro-Magnon people invented spear throwing.) Neanderthals instead thrust their long weapons directly into bison or woolly rhinoceros at close range, probably by ambushing them, jumping on their backs, then jamming the spear between their shoulder blades. This was nothing like going to the local grocery. (The hair on woolly mammoths and rhinoceroses could be several inches think and acted almost like armored plating.) If the thrust wasn't made instantly and accurately, being tossed like a rag doll and then gored would have been a very likely alternative outcome. No wonder their bodies resemble the battered torsos and limbs of broncobusters.

Personal sorrow aside, for Neanderthals every life lost must have been disastrous. Given their sparse populations, they didn't have many people to spare, nor, so far as we can tell, did they often live more than thirty years or so. Their productive years were pretty limited. Despite being spread out all across Europe and well into western Asia, genetic information gleaned from a handful of bones indicates that the total population of adult Neanderthals at most reached seventy thousand, and during the last forty thousand years of their existence probably dwindled to ten thousand, until finally they departed for good. Either way, dispersed as they were across tens of thousands of square miles, their clans couldn't have been large, probably smaller than bands of Native Americans that later roamed the western plains of North America for thousands of years.[4] Even meeting other clans must have been rare, and that would have left small groups, hardly more than extended families really, twelve, maybe as many as twenty-five people, fending entirely for themselves for long periods. Between injury, harsh weather, disease, and malnutrition, the whittling of their kind might have been slow, but it was, by all indications, also inexorable, and ultimately lethal.

Despite their rarity, Neanderthals survived a remarkably long time during a period when the climate was both harsh and unpredictable, fluctuating wildly sometimes within a generation or two, thanks, in part, to multiple volcanic eruptions around the planet. This longevity begs the big question, which is debated, passionately, among paleoanthropologists, exactly how complex was Neanderthal culture and how much did it have to do with their long-term survival?

At one end of the spectrum, some feel that they weren't much more advanced than the brutes Boule imagined in the early twentieth century—bereft of religion, language, much clothing, and any symbolic thought. Others speculate that they were as advanced as we were, or nearly so, with full command of some kind of language, poignant self-awareness, symbolic thoughts, and a rich social culture.

Where they fell along this spectrum probably has a lot to do with language. Without complex language, it's difficult to share and preserve ideas, whether they involve ceremonies, technologies, survival strategies, or relating what Aunt Marge has been up to. The ability to export an original thought from one mind to other minds has enormous advantages. Not only do good ideas spread rapidly this way, to the benefit of everyone who learns them, but it also increases the chances of the idea's remaining in the broader culture because more minds have glommed on to them. And once glommed onto, there is always the chance that someone else will improve it. That is one of the ways cultures form to begin with. Were the Neanderthals capable of this?

Maybe.

Steven Mithen, an archaeologist at Reading University in England, believes that early humans going back millions of years slowly developed a sense of rhythm, which was later combined with musical sounds that themselves became ways to communicate—soothing their children, winning mates, or motivating themselves. Later *Homo erectus*, and later still, he argues, Neanderthals, combined these primal musical skills with gesture and a kind of speech to develop a complex communication system he calls *hmmmm* for "holistic, multi-modal, manipulative, and musical."

It's not implausible. We humans are the only mammal, or primate for that matter, that can tap our feet in time to a rhythm. Powerful selective forces must have been behind the evolution of rhythm for it to become such a unique skill. Our speech is loaded with pauses, starts, and tonal inflections that in the hands of a first-rate orator have a powerful musical quality. Language without tone and inflection is flat, like that of a bad B-movie robot, devoid of feeling, and also, as a result, much of its intent and meaning. It's the music in our voices, something that scientists call prosody, that gives human language so much of its emotion, humor, and irony. It imbues speech with multiple levels of meaning, many of which we simply "get" without consciously realizing it, another indication that it evolved before words themselves.

Music is marvelously powerful. Think of the effect a national anthem, a favorite pop song, the climactic close to a Beethoven symphony, or just singing a song with friends can have. (How else can you explain karaoke?) It's not difficult to see how primal chants might have combined with dance and early rituals to become more complex, and more precise, ways to share feelings, emotions, and ideas once we had evolved brains capable of inventing them, and minds large enough to have need of them.

Neanderthals and *Homo sapiens*, remember, split from a common ancestor and went their separate ways for nearly two hundred thousand years before crossing paths again in Europe. It's quite possible that they wrought sophisticated, but entirely different, methods of communicating the thoughts on their considerable minds. Both we and Neanderthals carry the FOXP2 gene in our chromosomes, a snippet of DNA key to the development of speech (but not *the* language gene as some have characterized it; there is no language gene). Maybe both species built on a foundation of musical beats and sounds put to use by their common ancestor *Homo heidelbergensis*, but then, when they parted, evolved different ways to share their thinking. This happens with the color of fur and the shapes of appendages. Why not communication?

We know the direction we *Homo sapiens* ultimately took. We combined sound symbols—words—with a certain musicality—prosody—that created a commanding way to both conceptualize thoughts in our minds and then share them with others. This was one of the greatest innovations in all of nature, and it supercharged the growth of human culture.

Mithen imagines Neanderthals took a different path and evolved a complex combination of iconic gestures (think of the "crazy" gesture we use, an index finger twirling beside our head), songlike sounds to express emotions (more complex versions of the cooing and keening sounds we make), outright song and highly expressive dance movements (à la ballet and Broadway), all in concert to communicate on levels so intricate that they are beyond what we can even imagine.

These weren't muddled, caveman efforts to ape our *Homo sapiens* language, according to Mithen. He believes and makes a compelling argument that Neanderthals were musical and gestural virtuosos compared with us and the other human species that came before them. While we specialized in using our brains and vocal gifts as ways to deliver packets of symbols made of sound, Neanderthals evolved hyperrefined senses of sound, movement, and emotion.

One reason they may have evolved this way of communicating is because the physical structure of their skulls and throats developed differently from ours, partly because they were adapted to cold climates, partly by chance. Our heads and necks surround our vocal tracts, and our uniquely shaped skulls house a long, descended tongue uniquely gifted at forming vowels as in *see*, *saw*, and *sue*. Anthropologist Robert McCarthy believes Neanderthals simply couldn't make these sounds. To explore his theory, he created a synthesized computer model based on reconstructed Neanderthal vocal tracts developed by linguist Philip Lieberman from Brown University.[5]

The model pronounces the letter *e* the way a Neanderthal might have. The result is a sound that is never heard in our speech, a vowel that doesn't quite sound like an *e* or an *a* or an *i*, but something in between. McCarthy says this shows that Neanderthals could not speak what are known as "quantal" vowels. For us, these vowels provide subtle cues that help speakers with different-size vocal tracts understand one another because they enable a hearer to tune his ear in just such a way that he perceives the sound as it is meant to be heard, a little the way a radio tunes into the right frequency of a particular channel. With quantal vowels it's not simply the way we *say* the vowel, but also how we *hear* it. We learn how to tune in, apparently, when we begin babbling as babies. In fact, this may be one of the primary reasons we babble in infancy at all. We are not simply learning to make language, but learning how to listen to it, too.

You might not think this could matter much, losing a few vowel sounds. But McCarthy and Lieberman argue that if Neanderthals were unable to attune their voices and hearing to quantal vowels, it would have been impossible for them to distinguish between, for example, the words *bit* and *beat*. Instead the Neanderthal *e* would have been substituted in both. The effect is that they would have had fewer vowels and therefore far fewer words to express ideas, and this would have encouraged our cousin humans to instead build on the more ancient *hmmmm* approach that Mithen imagines. After all, why would they have created words like *bit* and *beat* if they had little use for them in the first place? They would have had no experience of quantal vowels and no more reason to use them than we would re-create the alien speech of a Martian whose body and throat is shaped altogether differently from ours.

If McCarthy and Mithen are correct, perhaps Neanderthals compensated for a shortage of vowels with an abundance of tones. Maybe the vocabulary, as in Chinese, was based more on inflection and context, less on diphthongs, vowels, and consonants and the symbols these combinations of sounds could represent.

Neanderthals might also have failed to develop the verbal palette we did because their social world was smaller. Our communication is what makes social interaction rich, and that makes our own mental and emotional lives more elaborate. A less elastic way to express ideas might equate to fewer nuanced ideas emerging from the minds of our big, northern brethren. Could Michelangelo have hoped to paint the richly colored moment of creation on the ceiling of the Sistine Chapel if his palette consisted only of gray, black, and white with a little red? Maybe. We *are* talking Michelangelo. But it would have been an entirely different image altogether.

Symbolic language, and more specifically the spoken word, also makes logic more likely. It doesn't only enable us to store and communicate ideas and thoughts; it shapes and refines the sorts of ideas and thoughts we have in the first place. Without more refined language, maybe the Neanderthals' worldview was less logical, and more dreamlike, almost surrealistic. When we dream, the dream always makes sense within the context of the dream, even though when we recall dreams after we awake, we know that flying, and time travel, and the different versions of ourselves and others in the dream aren't

possible in the "real" world. Did the daily awakened life of Neanderthals possess more of this kind of ephemeral, nearly surrealistic quality?
We ourselves touch on the mystical, the surrealistic, and the metaphys-
ical in meditation, religious trances, and hypnosis. And doesn't dance
and music sometimes bring people to a mystical, trancelike state? If we
can come to these states, perhaps Neanderthals did, too, and then
some, given the way they perceived the world. It makes you wonder
what Neanderthal dreams might have been like. And it makes you
wonder how accurate our "reality" is.

Not that any of this means that Neanderthals were less adept at the
undertakings necessary to their survival. Mithen believes that they had
plenty of "domain specific" intelligence characterized by the tools
they made, the hunts they organized, and the food they prepared, but
there is little evidence so far that they wove ideas into a broad culture
filled with story and myth.

Their small numbers might have hindered their proclivities and talents. There couldn't have been much cross-pollination of ideas among
separate groups. Mithen even wonders if they developed hmmmm
"dialects" specific to their own clans. Each group would have been
like an out-of-the way island, rarely found. Given their dialects and
the rarity of chance meetings, technical and social progress would
have been stunted. In the long run that would have made the rise of
sophisticated culture difficult.

This may explain why the more closely anthropologists have explored Neanderthal culture, the more they have noticed something
odd. For all of their dogged courage and resilience, they didn't make
much technological progress during their two-hundred-thousand-
year run in Europe. The Mousterian tools and cultural artifacts they
crafted and left behind show remarkably little innovation considering
how long they were around. The craftsmanship is first-rate, and the
design of the tools and the methods used to fashion them were clearly
passed along quite precisely, but given their intelligence, you would
have expected more novelty, more originality.

Their greatest technical breakthroughs seem to have come *after*
they first crossed paths with the Cro-Magnon people. This could be
coincidence or it could illustrate what they might have accomplished
if they had been able to better share ideas among themselves. Or it
could mean their sparse numbers and the limitations of their lan-

guage hampered the two species' ability to interact once they finally and fatefully met.

Imagine this encounter, and its shattering effect. Each group must have gazed at the other in bewildered amazement. In an instant they would have seen that these creatures resembled them, but were clearly not one of them. Why didn't they communicate in the same way or even make the same sounds? This wasn't simply a different tribe that dressed in unfamiliar apparel, spoke an indecipherable language, or carried odd weapons. This was another creature altogether, perhaps a god or an animal or something in between. To the Cro-Magnon (see sidebar p. 114), the large-muscled, beetle-browed white people with their fiery hair must have struck them as alien, and possibly dangerous. To the broad-backed Neanderthals, the slim creatures with their baby faces and rounded skulls might have looked slight, childlike, and at first glance weak. But the Neanderthals may have sensed danger, too, in the sophisticated weapons the strangers carried, and in the alien precision of their communication. Chances are the Neanderthals had seen the handiwork of those weapons long before they met face-to-face the creatures who had fashioned them. The evidence of such efficient killing must have had a chilling effect.

The big and primal question—the mastodon in the room so to speak—that had to have entered both of their collective minds was, whoever they are, can they be trusted? Are they a friend or an enemy?

For twenty-five thousand years, nearly three times longer than we have been recording our own history, *Homo sapiens* and Neanderthals shared the same part of the world. Over time, and as the Cro-Magnon people wandered deeper into Europe, the species must have met again and again. Did they cooperate or wage war or simply do their level best to ignore one another while each worked desperately to stay beyond death's long reach?

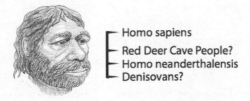

Homo sapiens
Red Deer Cave People?
Homo neanderthalensis
Denisovans?

Who Were the Cro-Magnon People?

The term *Cro-Magnon* can be a little confusing because it origi-
nates from a French cave, Abri de Cro-Magnon, in southwestern
France, where the first fossils were found, but actually refers to
the dark-skinned people whose ancestors began migrating from
Africa around fifty thousand years ago. These were the earliest
Homo sapiens to reach Western Europe, and the people who first
encountered and then coexisted with Neanderthals in places we
now know as France and Spain beginning some forty thousand
years ago.

As you might imagine, the Cro-Magnon were a tough group.
Strong, heavily muscled, and smart with a brain, at 1600 cc, larger
than ours is today. Their tall foreheads and square jaws made them
the first humans (as far as we know) to bring these neotenic features
with them into their adulthood, the physical hallmarks of modern
Homo sapiens.

They were clearly successful. Their genes are evident in people
living today from Europe to Central Asia and North Africa, Polynesia and both American continents. In short, nearly all of us.
Their weaponry was advanced and included the invention of bone
spear throwers that held their spears as they launched them at prey
(and likely one another on occasion) with a force and accuracy that
made them the most lethal hunters on earth. They also excelled in
fashioning extremely sharp flint knife blades and spearheads. They
even developed techniques for straightening their spears to make
their flight more true. They liked to decorate their weapons, too,
but discoveries of these small examples of their flair for the artistic
were only a small indication of their ingenuity. The world learned
of the true depth of their creative talents in 1940 when four curious teenage boys, and their dog, Robot, stumbled upon arrays of
mysterious paintings on the walls of the Lascaux caves in France's
Dordogne region.

The artwork is nothing short of jaw-dropping, as beautiful and
haunting as anything a modern artist could possibly conjure, and
an indication that in these people, modern human behavior had
irrevocably touched the world. Why the paintings were created is
unknowable. They may have been religious, or a way to enter a
spirit world, or simply the doodlings and artistry of generations of

ancient but extraordinary humans who were expressing themselves in ways other humans never had. Hundreds of caves have
now been discovered around the world filled with the imaginings
of these and other ancient humans, all of them powerful illustra-
tions of the playfulness and creativity, the childlike side of us, that
distinguishes our species.

As thoroughly as the archaeological record has been pored over, it
has yielded nothing more than the skimpiest portrait of how the
two species lived, let alone how they may have interacted. *Homo sapiens*, we know, had set up trading relationships with one another
thousands of years earlier, which increased cooperation and improved
their chances of survival. But because both species were itinerant,
neither had yet established villages or cities, though there were settlements, favored places that clans and bands regularly returned to over
long periods.

Neanderthal and early Cro-Magnon existence may not have been
terribly different from the way Native Americans lived on the plains
of North America as recently as the nineteenth century—moving
with the seasons, following the large animals they fed on, hunkering
down in the winter against the elements, in and near caves that provided warmth and shelter, and then moving again when the weather
grew a little kinder. Generation after generation they likely lived this
way, bending with the climate, following the herds of mammoth, elk,
and deer that provided them with food, clothing, bones for tools,
many of the raw materials they relied on for their existence. Life was
a short, harsh cycle of perhaps thirty to thirty-five winters and sum-
mers of close cooperation, family feuds, and occasional encounters
with other humans, then death, at which time the next generation
took on the fight. In some ways it isn't all that different from our existence today. Life was shorter, it's true, and tougher and the technologies different, but the same general pattern applied. They too sought
love, enjoyment, and friendship and searched for ways to express
themselves, just as we do. They were, after all, human as well. Just a
different variety of human.

All of this only makes it more tempting to wonder what happened during those long and wintry twenty-five millennia when
our kind and Neanderthals coexisted. Why did Neanderthals fail
to survive? It's a vexing mystery. Every species runs its course.

We know that. And the Neanderthal had made an immensely successful run. They roamed the steppes and mountain forests of Europe and western Asia through three ice ages, and their close cousin *Homo heidelbergensis* had survived a full two hundred thousand years before departing. During most of their time the Neanderthals were the dominant primate species north of Africa. But as the last glacial age began, ever so slightly, to wane, perhaps for the Neanderthal people their time for departure had simply come just as it had come for so many others before them. If that was the case, the arrival of modern humans couldn't have helped their situation whatever the intentions of *Homo sapiens.* The Cro-Magnon were moving into the Neanderthals' ecological niche and were proving to be better survivors.

Some have speculated that we systematically wiped out our long-lost cousins as we came across them. When we met, the theory goes, if hunting or choice settlements and locations were at stake, the Cro-Magnon, with their superior weapons, and possibly their superior planning, killed or enslaved whoever got in their way, including Neanderthals. (They might have done the same to their own kind. We still do today.) It wouldn't have been an all-out war in the sense that armies were assembled and clashed, but the damage done to the Neanderthals would have been relentless, with one settlement, tribe, or clan after another falling to the new intruders.

There's not much evidence for warfare or murder in the fossil record, however. We haven't found ancient killing fields, strewn with the hacked and broken bones of the two species; no sites where dead *Homo sapiens* lie next to the skeletons of Neanderthals. The first evidence of a violent Neanderthal death was discovered in the Shandihar Cave in northeastern Iraq. The man was about forty years old when he met his end. Scientists found evidence of a wound from a spear, or some sort of sharp object, in his rib cage. Based on the nature of the wound, Steven Churchill, an anthropologist at Duke University, suspects a light spear thrown by a Cro-Magnon enemy inflicted it. It doesn't seem to be the result of a thrust by a knife or a long Neanderthal spear, the kind they favored when hunting. It's a theory, but a long way from a certainty. If this one man was murdered in this cave, or nearby where were the other victims? It's just as likely that the man died from a wound suffered while hunting, or maybe another Neanderthal did him in.

Other findings have been a little more conclusive, and considerably more gruesome. Paleontologist Fernando Ramirez Rozzi found something rare in the human fossil world, a cave called Les Rois in southwestern France that housed the bones of a modern human and a Neanderthal child lying together, positive proof that the two species came face-to-face. Unfortunately the jawbone of the child shows the same sort of markings paleontologists see on the bones of butchered reindeer skulls. The unappetizing conclusion is the child was made a meal of. Hints that other Neanderthals met a similar, cannibalized fate have been found at a site called Moula-Guercy near France's Rhône River. Except in this case those who dined on their fellow humans were themselves Neanderthals. Perhaps it was a violent ritual, or the spoils of war, or maybe some who had died of starvation became the sustenance for those who survived. Not a cheery thought, but a world this harsh would inevitably require harsh choices.

If there were violent meetings, then this is the extent of the evidence we have for now. Others if they exist have yet to reveal themselves. Maybe, somewhere in Europe, in a remote mountain forest or beneath a broad river rerouted by the last glaciers, lie the bones of prehistoric warriors who fell to the invaders from the southern seas. So far, though, no battlefields, and no warriors, have been found.

A second theory that could explain the disappearance of the Neanderthals is that Cro-Magnon simply outcompeted them for resources, food, and land, not unlike the way we outcompete nearly every other species living wherever we show up. The thinking is that we didn't kill them hand to hand, but we exterminated them in a war of attrition, by taking over the best habitats and hunting grounds, killing game faster than they could, and in larger numbers. Slowly, over thousands of years, the already sparse Neanderthal population retreated to pockets where it became increasingly difficult for them to survive. (We are doing this today to the gorillas and chimpanzees of Africa and the orangutans of Southeast Asia.) This might further have crippled Neanderthals' ability to band together, weakening them still more, until in the end each of the dwindling clans died away.

There is some evidence for this. Neanderthals did become progressively rarer as Europe withdrew into the coldest phase of the last ice age. Leslie Aiello of University College London suggests that Neanderthals, as adapted as they were to chilly climates, couldn't survive temperatures below 0° F (−18° C). Their clothing and technology simply weren't up to it thirty thousand years ago. As temperatures dropped and a new ice age descended, warm pockets of land would have become increasingly hard to find. If the Neanderthals retreated to them, they may have been trapped and died as even these locations grew too cold. Or they may have sought them only to find that the new creatures from the south had beat them to it, leaving them bereft of their favorite settlements and with no choice but to settle for places that, in the end, couldn't sustain them.

It might have happened this way. But Europe and Asia are immense territories, and it's difficult to imagine that there wouldn't have been enough resources to go around. The Neanderthal range covered tens of thousands of square miles. Genetic studies indicate the entire Neanderthal populations rarely numbered more than seventy thousand people spread from the Iberian Peninsula and the south of England clear to the plains of western Asia beyond the Caspian Sea. While each band probably needed several square miles of land to sustain them, much of the land was rich with food and resources and herds of large animals from mammoths and woolly hippopotamuses to deer, bison, and aurochs.* Even if the combined numbers of both Neanderthals and modern humans reached into the hundreds of thousands, there would seem to have been plenty of space, food, and resources, and much of Europe, even at the height of the last ice age, would have been temperate enough to accommodate any variety of human— Neanderthal or not. The frigid weather would certainly have battered Neanderthals trapped in cold areas, but why wouldn't those already living in the more temperate climates of southern Italy, Spain, France, and the Mideast have survived?

Maybe because it was more complicated than any one of these scenarios. Maybe the mysterious people from the south brought new diseases or parasites with them or forced radical cultural changes that Neanderthals simply couldn't adjust to. After all, white men from Europe destroyed the cultures and ways of an entire continent of native North Americans, scores of individual tribes numbering in the hundreds of thousands, and they did it inside of four hundred years. This

---

*Aurochs were a type of now-extinct, large, wild cattle that inhabited Europe, Asia, and North Africa. They survived in Europe until 1627, when the last recorded member of the species, a female, died in the Jaktorów Forest, in Poland.

wasn't simply a matter of brute slaughter. The blunt impact of a different kind of culture can also do considerable damage. Could immigrants from Africa have wreaked the same kind of havoc on the Neanderthal natives of Europe, except in this case taking twenty-five thousand rather than four hundred years? It's possible.

Stephen Kuhn and Mary Stiner at the University of Arizona suspect that modern humans arrived in Europe with cultures that divided labor within their tribes in a way that was safer for pregnant women, mothers, and children by keeping them focused mostly on collecting vegetables, fruits, and nuts, while men concentrated on hunting large animals. Based on their research, Kuhn and Stiner believe Neanderthals divided their labor among the sexes differently or, more accurately, didn't divide it at all. Men and women both undertook the deadly work of bringing down big game, and that meant that women who were killed in hunts would not survive to bear more children. The teens and adolescents lost in hunts would further have depleted the clan.

Though the Cro-Magnon approach to dividing labor didn't mean they attacked Neanderthals, it would have had an impact nevertheless because eventually more Cro-Magnon women would have survived to bear more children than their Neanderthal counterparts, growing their population while Neanderthals struggled to keep pace with replacing the members they were losing. Even if the Neanderthal people were tougher, over thousands of years the competitive difference could have completely rearranged the population balance, just as divergent social approaches shifted the balance between whites and Native Americans.

This may help explain why the Neanderthal population never really took off, even when ice ages relented. Their mortality rate was simply too high, and they were spread out too thinly. It may also help explain why their culture and technology remained doggedly unchanged for two hundred millennia. It's terribly difficult, even within a clan, to pass along new ideas and innovations when members rarely lived past thirty or thirty-five years, and others are being wiped out in the prime of their childbearing years. Who knows how many Neanderthal Galileos or Einsteins died suddenly in the hunt and took their genius and inventions to the grave with them? It's nearly impossible to build anything but the most rudimentary traditions when innovation is rare and life passes so quickly. In this scenario, the Neanderthal found themselves fighting, millennia after millennia, a pulverizing war of attrition. In the end, extinction was the only possible outcome.

One last theory about the demise of Neanderthals is particularly tantalizing: If we killed them at all, we killed them with kindness. We neither murdered them nor outcompeted them. We mated with them and, in time, simply folded them into our species until they disappeared, reuniting the two branches of the human family that had parted ways in Africa two hundred and fifty thousand years earlier when small groups of restive *Homo heidelbergensis* headed across North Africa and into Europe.

It's fascinating to consider the possibility that we and another kind of human together bred a new version of the species. Whether this happened was, only a few years ago, one of the great controversies in paleoanthropology. But now there is persuasive evidence that something like it did.

In 1952 the remains of an adult woman were found lying on the floor of the Pestera Muierii cave in Romania—a leg bone, a cranium, a shoulder blade, and a few other fragments. The people who discovered these bones didn't think much of them. How old could they be, after all, if they were simply lying there on the ground for anyone to kick around? As a result, soon after their discovery they were squirreled away in a researcher's drawer where for more than half a century they lay undisturbed and forgotten. Eventually, however, a team of scientists that included Erik Trinkaus at Washington University in the United States and two Romanian anthropologists, Andrei Soficaru and Adrian Dobos, rediscovered the bones and gave them a closer look, and when they did, they were stunned. Radiocarbon dating revealed the woman hadn't lived recently at all, but last walked Earth thirty thousand years ago. The other startling discovery was that the fossils exhibited features that were clearly Cro-Magnon-like, but also distinctly Neanderthal. The back of the woman's head, for example, protruded with an occipital bun, a distinct Neanderthal trait. Her chin was also larger and her brow more sloped than a modern human's. The woman's shoulder blade was narrow, not as broad as a modern human's. Was she simply a rugged-looking modern human, or, as one scientist wryly put it, proof that moderns "were up to no good with Neanderthal women behind boulders on the tundra?"

Other similar finds made recently throughout Europe keep boggling the minds of scientists who study this question. In another cave in France researchers have unearthed not bones, but tools that date back thirty-five thousand years. Their location indicates that for at least a full millennium both Cro-Magnon and Neanderthals coexisted in this place. If they could live together, and if they could communicate and cooperate, isn't it likely that at least a few crossed the species line and, in a prehistoric foreshadowing of Romeo and Juliet, mated?

Then there is the mysterious skeleton of a young boy unearthed in Portugal that is 24,500 years old. While conventional wisdom has it that the last Neanderthals died out thirty thousand years ago, the large size of this boy's jaw and front teeth, his foreshortened legs, and broad chest have caused Trinkaus and others to wonder if he, too, might not be a hybrid. Though his chin is Neanderthal in size, it is also square, more like ours, and his lower arms were shorter and smaller than you might expect if he were *Homo sapiens*.

Strangely enough, this part of Portugal is among the last places in Europe where Neanderthals lived before they disappear from the fossil record. Was this boy simply among the last of his kind, archaeological proof that Neanderthals were finally and inevitably swallowed, genetically or otherwise, into the rising tide of modern humans spreading across the planet?

Until recently, the only evidences of interbreeding were perplexing finds like these, smoking guns that indicated we and Neanderthals had mated, but nothing irrefutable. Then in 2010 a scientific consortium headed by the Max Planck Institute for Evolutionary Anthropology completed its historic analysis of the Neanderthal genome, accomplishing for our burly cousin species what we had done for our ourselves seven years earlier. The analyzed DNA was extracted from three Neanderthal bones discovered at the Vindija Cave in Croatia not far from the Adriatic seacoast. To decipher the tantalizing possibility that we and Neanderthals may have produced common offspring in the deep past, the team compared the Neanderthal DNA with the genomes of five people of different lineages from around the world— French, Han Chinese, Papuans from New Guinea, and the Yoruba and San people of Africa. The San are, genetically, very close to the first modern humans to have evolved in Africa.

What dumbfounded the project's investigators, and the rest of the scientific world, was that all the genetic samples taken, except for the

Yoruba and San people of Africa, contained 1 to 4 percent Neanderthal DNA. In other words, most of the human race from Europe to the islands of Southeast Asia (and probably farther) is part Neanderthal! That Africans seem not to share any Neanderthal blood indicates that these two families mated after the wave of *Homo sapiens* departed Africa, but before their descendants headed into Europe and Asia. According to the researchers, this would have been somewhere between eighty thousand and fifty thousand years ago.

Was that the only time modern humans and Neanderthals bred? The research team isn't saying, but right now they can only base their conclusions on the research in hand. This will disappoint those who believe that Neanderthals and modern humans melded during those twenty-five thousand years of cohabitation in Europe into a single species whose recombined genes, shaped by separate evolutionary pressures, created a new kind of human. But it doesn't rule the possibility out. There simply isn't enough information on the scientific table right now to say.

That we mated still doesn't conclusively solve the mystery of how the hardy, quiet people of the North met their end. Was it murder, competition, or love? Does it have to be one or the other? Nature, evolution, and human relations are all chaotic and unpredictable, as much as we might like them otherwise. When Europeans colonized North and South America, they sometimes befriended the natives, sometimes brutally exterminated them, sometimes raped their women, and sometimes fell in love and raised families. Were Neanderthals so different from Cro-Magnon that sex was out of the question? If the Max Planck findings are accurate, clearly not. Both species were human, and the drive to procreate is strong and primal. Humans, after all, have been known to have sex with other primates, even other animals. Surely both species found enough common ground over twenty-five thousand years to bed down together during those frigid European winters. One of the inescapable lessons of evolution is, if anything can happen, it probably will.

Whatever ultimately transpired between our two species, events eventually rendered Neanderthals first endangered and finally extinct. For two hundred thousand years they were punished by a cold climate that kept their numbers small and made it difficult to develop stable trade, and to share and amplify the advances of their scattered clan cultures. Even with complex forms of communication, no matter how different from ours, they would have been hard-pressed to build a broad and increasingly sophisticated culture. They seem to have had their work cut out for them simply maintaining the status quo. So, in time, they disappeared, presumably one band at a time. Sometimes, perhaps, at the hands of one another. Sometimes to disease or famine. Sometimes to climate change, undoubtedly a particularly destructive culprit, especially if their technologies couldn't keep up as modern humans began to usurp their favorite places to live, killing game before they could and depleting limited resources.

Sometimes the Neanderthal may have battled for those places and lost or moved on, as Native Americans did until nothing but the poorest lands were left upon which to scrape out a life. Or sometimes they may have compromised and cooperated and befriended their clever, slim competitors, until there was no difference between the two, and they had disappeared into the gene pool of their cousin species, leaving a few bones and tools and a bit of their DNA in us as a legacy of their more prosperous days on earth.

Precisely how the end came is impossible to say, but inevitably, somewhere, sometime, the last Neanderthal passed from this earth. Current theories hold that surviving bands may have retreated south during the last ice age to the Iberian Peninsula, holing up on Gibraltar, a last outpost at the toe of Western Europe.

Gibraltar protrudes like a great snaggled tooth from the southern tip of Spain no more than fourteen miles from the north-facing coast of Africa. For one hundred thousand years Neanderthals had been revisiting and living in a great cave, today known as Gorham's, that sits at the base of the peninsula's massive promontory rocks as they hover hawklike over the Mediterranean. If the radiocarbon dating of nuggets of charcoal found in the cave is accurate, the Neanderthal stoked their last fires here twenty-four thousand years ago, much more recently than scientists once thought possible.

Gibraltar was different during the last ice age than it is today. Populated with deer and ibex and rabbits in the craggy hills above the cave's colossal and vaulted ceilings, it made a perfect refuge for a vanishing species. With sea level in the Strait of Gibraltar 240 to 360 feet lower than it is today, the cave would have gazed like a cyclopean eye on sprawling plains and marshlands that stretched west across the great bay rather than the blue sea that surround the land now.

The bones left behind tell us the cave dwellers dined sumptuously on tortoises, fish, and other marine life. It had long been a good place to be, and now, it seems, it was the *last* place to be. For the Neanderthal, only Africa itself could take them farther south. To the north the Cro-Magnon had been moving into their territory for millennia, and beginning six thousand years earlier the climate had grown fiercely colder, driving clans south until finally only this one last settlement remained, clinging to the underbelly of the continent.

Gibraltar, surrounded by water and so far south, would have remained temperate even in the face of the descending glaciers, up to a point. By twenty-four thousand years ago, as the ice age tightened its frigid grip, even Gibraltar grew arid, the marshlands died off, and the game with it. Each day would have become a little harder, and finally, impossible.

Someone had to be the last Neanderthal, an individual like you and me. He, or she, didn't know it, but whittled down and alone, that person's end was more than a single death. It was the passing of an entire species shaped and hammered in evolution's crucible for hundreds of thousands of years. What were those last hours like?

I would like to think they were spent sitting on a Gibraltar precipice, high above the shallow Mediterranean, looking west as the sun descended into the Spanish mountains. Maybe in those last moments the pale light faded on the Neanderthal's sloped and beetled brow while that strange and fiery orange ball slipped mysteriously away, and with it the last Neanderthal mind, with its last, unique Neanderthal thoughts. After two hundred thousand years, a time that mangles and boggles our ken, extinction had at last come.

But why the Neanderthal and not us? To say it was because we were smarter somehow, or better communicators, or more social, strategic, or creative begs the deeper question, what happened that we developed those gifts and not them? What made the difference? As usual, the answer isn't simple, but it's fascinating, and once again it is linked to our childhood.

When teeth are formed, they are built one layer of enamel at a time, each deposited on top of the other. This leaves a pattern, if you inspect them closely enough, that resembles the growth rings of a tree; the more enamel that is laid down, the wider the spaces, or perikymata, become.

You wouldn't think that ancient teeth could possibly have much to say about human evolution, but, as it turns out they speak volumes and provide a terrifically convenient way to measure how quickly different primates, including our human relatives, grew up. Even if a scientist has nothing more than a single, ancient molar or bicuspid to work with—which, in paleoanthropology, is often the case—it's remarkable how illuminating a tooth can be. By inspecting the speed with which the perikymata of even very different primates were laid down, it is possible to compare how "old" they were, relative to one another, when they reached adulthood, or puberty, or even when they were likely weaned, all depending on when the tooth stopped growing. In this way a tooth can act as a rock-solid biological clock set against other biological life events that can help shed some much-needed light on whether our ancestors matured more quickly, or less quickly, than we did.

You and I, we already know, take eighteen to twenty years to reach physical adulthood, whereas chimpanzees and gorillas reach adulthood by age eleven or twelve, in nearly half the time. The main reason for this is our extended childhood. The point is, our rates of growth over our lifetime are different from those of other primates. We cut our first permanent molars around age six, but chimps lose their baby teeth around the age of three and a half. The earliest humans, such as *afarensis*, developed at the same rate as chimpanzees, and so did their teeth. But later, with the arrival of *Homo habilis* and *Homo erectus*, childhood lasted longer and growth slowed. A *Homo erectus* child cut her first molar between ages four and four and a half.

All of these findings pooled together from studying the teeth of precursors from around the world readjusted the age of the most famous youngster in all of anthropology, the Nariokotome (or Turkana) Boy (see chapter 2, "The Invention of Childhood"). Originally scientists pegged his time of death at age twelve, but now the consensus is that he was closer to eight even though he was already an impressive five feet three inches tall. He was hitting his adolescent growth spurt (the one that drives parents crazy when they are trying keep their kids in reasonably fitted clothes) even though by our lights he should still have been a little boy. (He was, therefore, about halfway between us and chimpanzees in his growth rate.) All of this tells us that earlier in our evolution our ancestors grew up faster, which means their childhoods were shorter, which further means that they had less time to learn before they began to get set in their adult ways.

You would think that by the time Neanderthals had arrived on the scene that the speedier growth rates of Nariokotome Boy would have evolved out of us and caught up to rates similar to ours today. After all, we and Neanderthals evolved from the same common stock and came into existence at about the same time. We were roughly the same physical size and so were our brains.

For some time it looked exactly this way, then in 2001 Alan Walker, one of the team who had originally discovered Nariokotome Boy, found that Neanderthals didn't attain modern growth rates like *Homo sapiens* until about 120,000 years ago, 80,000 years after they first arrived. Or so he thought. But soon this conclusion was proven wrong when a Harvard researcher named Tanya Smith and her colleagues, after the careful inspection of many teeth, concluded that not only did Neanderthals not lengthen their lives or their childhoods as much or as early as Walker had thought, they had actually shortened them, reversing a trend at least seven million years in the making! Smith says Neanderthals reached full maturity by age fifteen, three to five years earlier than us and not terribly different from the pace that Nariokotome Boy was on, nearly a million and a half years earlier. On the other hand, human fossils unearthed in Morocco indicate that we *Homo sapiens* reached current growth rates as early as 160,000 years ago.

What, however, would cause a childhood-lengthening trend that had been in the works for so long to reverse itself in Neanderthals? The same force that causes all evolutionary trends to turn and twist —the need to survive.

Neanderthals, you might recall, had a rough time of it fighting cold climates and hunting enormous animals at close quarters, among other challenges. Their population, even when the climate grew warmer, never took off, which meant that from the first moment they emerged, they were, essentially, an endangered species. It's true they were enormously strong and as tough a creature as ever walked the planet, yet they didn't live long. Because they were so quickly snuffed out, and because they congregated in small groups, evolution apparently began to favor Neanderthal children who grew up faster, could bear children sooner, and reached adult size and strength as rapidly as possible to replace the older members of the troop who passed on so quickly.

This would have two immense and not terribly favorable long-term effects. First, it meant Neanderthal children spent less time playing, learning, and developing socially and creatively in early life. The effect—less personal adaptability and creativity. They had less time to develop unique personalities and talents. Second, it meant fewer mentors who could pass valuable knowledge along to younger members of the clan. Neanderthals became so focused on their short-term need to survive that they were unable to develop the more complex skills that saved us *Homo sapiens* over the long haul.

From an evolutionary point of view, however, what other route could they have taken? Neanderthal mothers could not suddenly begin to have litters of offspring like a cat or a pig to compensate for their kind's high mortality rate, and given their sparse numbers and tiny tribes, they simply didn't have the "bench strength" that a larger population supplies. They were caught in an evolutionary catch-22, and accelerating their childhoods was the best Darwinian solution at hand. For two hundred thousand years, it worked. And then it didn't.

We *Homo sapiens* were luckier. Though we had swung precariously close to extinction ourselves fifty thousand years earlier, the climatic forces behind our demise struck quickly, then reversed. Despite our near-death experience, there wasn't enough time for a genetic solution à la the Neanderthals, so when the climate recovered, so did we, in a hurry. DNA analysis shows we rapidly fanned out into Europe and Asia, and all points beyond. The main reason we could was because we had already maximized the lengths of our childhoods, which now poised these strange, slender savanna apes with their youthful looks, big brains, and enormous personal diversity to change the world in profound and startling ways.

Which brings us to the next part of our story.

CHAPTER SEVEN

BEAUTIES IN THE BEAST

I cannot imagine, even in our most primitive time,
the emergence of talented painters to make cave paintings without there
having been, near at hand, equally creative people making song. It is, like
speech, a dominant aspect of human biology.
—Lewis Thomas, *Lives of a Cell*

YOU MAY NOT find it particularly attractive to perforate your upper lip and then slip a large metal and bamboo ring called a pelele into it to force your lip two inches beyond your nose, but women in the Makololo tribe of south-central Africa did it in the nineteenth century, and the men loved it, even when a smile sent the ringed lip flipping up to cover the eyes of the woman who wore it.

In 1860, when a British explorer asked the Makololo chief, "Why do women wear these things?" the chief, in stunned disbelief, answered, "For beauty! They are the only beautiful things women have; men have beards, women have none. What kind of a person would she be without pelele? She would not be a woman at all with a mouth like a man, but no beard."[1]

It is difficult to underestimate the power of the visual cues that drive human behavior, including those devoted to the arts of seduction. "Savages at the present day deck themselves with plumes, necklaces, armlets, ear-rings, etc. They paint themselves in the most diversified manner," Charles Darwin wrote in 1860. He devoted an entire chapter in *Descent of Man* to exhaustively detailing the wild and alien ways people all over the world embellished themselves to attract and impress the opposite sex. The natives of Malaysia painted their teeth black because it was shameful to have white teeth "like those of a dog."

Some Arabs believed that no beauty could be perfect until the cheeks or temples "have been gashed." The Botocudos of Brazil placed a four-inch disk of wood in their lower lip, and the women of Tibet elongated their necks by placing metal ringlets one on top of the other until their heads almost appeared to hover magically above their shoulders.

Most of the time, Darwin observed, women's adornments focused on enhancing their beauty. Men concentrated on making themselves physically attractive, too, but mostly they favored adornments designed to strike terror into their enemies during battle because a fierce warrior is often attractive to the opposite sex. So they embellished themselves with paint of all kinds, or, like the Maori of New Zealand with remarkably detailed facial tattoos. The women of some African tribes found a star stamped on a man's forehead and chin absolutely irresistible.

Darwin's anecdotes aren't the only testament to how important we consider our appearance to be. Human beings focus on appearance all the time, everywhere. In 2011 the cosmetics industry induced men and women worldwide to separate their wallets from $12.5 billion. And in 2010, Americans, without help from anyone else in the world, spent $50 billion on jewelry.[2]

Darwin enumerated example after enthralling example of this aspect of human behavior because he was trying to make *the* central point of *The Descent of Man*—species work to guarantee their survival in two ways. First, by outflanking disease, parasites, predators, foul weather, and all the other countless dangers of their environment. And second, by having sex. Only by finding willing mates with whom to get on with the business of bringing new offspring into the world, he pointed out, can any species hope to survive. This process he called "sexual selection." The two strategies are intimately bound. Survival serves no purpose without sex, and sex, of course, is impossible without survival. Naturally, the first step to being sexually selected is getting the attention of the opposite sex in the first place. You can't mate if you can't manage to be irresistible. To go unnoticed is to go unloved, and unrequited love in nature is a sure path to extinction, at least for you and the DNA you personally have to offer the gene pool.

This makes fewer goals in life more important—from an evolutionary point of view—than successfully landing at least one sexual

A Hen Is an Egg's Way of
Making Another Egg

Being self-aware as we are, we tend to think that the drive to survive is a conscious thing, and so we assume that this awareness of our own mortality makes us want to remain living. But every form of life—the lowliest protozoan, deep-sea tube worm, or hardy lichen clinging to a windswept antarctic rock—fights every day, ferociously, to remain among the living. Lizards, spiders, gazelles, and lions, all focus themselves ardently to the quotidian labor of making it to the next day, yet not one of them is contemplating its mortality. The drive to live is instinctual, primal, and unconscious, even in us. But where, and this is the central question, does the instinct come from?

You couldn't be blamed if you assumed it comes from the individual living thing itself, but again, so much of life doesn't have the cerebral horsepower to even know that death is possible. So something else must be at work, and it is. Long ago, packets of molecules with the remarkable ability to continually make copies of themselves evolved. Scientists and author Richard Dawkins like to call these "survival machines." In time these evolved into what we now call DNA, the long ladders of linked proteins that contain the instructions that make you and me, and every other living thing on the planet, rather improbably and astoundingly possible.

To better do their work, the earliest DNA replicators inevitably stumbled across ways to better multiply. The very first cells are an outstanding example of a major leap forward. They not only supplied a membrane as a protective wall between them and the cruel protean world, but they discovered ways to ingest food and turn it into power, the better to make even more copies. Sex was another innovation—a better way to make both more and more diverse survival machines. In time cells joined together to form increasingly complex replicators, until following 3.8 billion years of trial and error, they took on millions of outrageously complex forms. One recent, and altogether serendipitous, result is you.

The British poet Samuel Butler once observed, "A hen is only an egg's way of making another egg." Not the other way around.

When looked at this way, it turns out you and I (and every other living thing on earth) are not so much focused on surviving because we personally want to avoid death or even desire to create more versions of ourselves. Instead we are a kind of elaborate tool in the unconscious service of the DNA swimming around inside us, determined (if strings of molecules can be determined) to make more copies of itself. Think about that. We are hosts to a kind of virus that controls our fundamental behavior in a way that ensures more copies of that "virus" will be made because that is what that virus does—it replicates. And the better "tricks" it can find that improve its duplication, the better it does its job. We, in case it escaped you, are one of the "tricks."

partner. This has caused the forces of natural selection, given their collective knack for conjuring strange genetic fabrications in the interest of survival, to cook up some extravagant ways to advertise just how alluring the members of various species can be to one another. Peacock feathers are surely the most celebrated, but there are also the colossal antlers of the (now extinct) Irish elk, the elaborate songs that the red-eyed vireo sings to charm females, the thick manes of lions, and the vibrantly colored bottoms and faces of male mandrills. Even the bright colors and fragrances of flowers are a kind of sexual allurement because they attract bees that then "impregnate" other flowers by proxy.

What makes so many of these evolutionary traits intriguing is how extravagant they are. They don't seem to serve any practical purpose, not at first glance. In fact they can sometimes get in the way of survival because they require enormous strength, extra stores of nutrition, or draw the attention of predators. Biologists call these costly attention getters "fitness indicators" because they are billboards that mostly male animals use to advertise to females the fabulous genes they are toting around. Extravagant dances, battles with other males, wildly orchestrated warbling, gargantuan antlers, luminous bottoms, thick manes—all of these are powerful, but costly, signals to the opposite sex. Their sole purpose is to prove "I am the man!"

Like other animals, we humans have evolved an impressive variety of fitness indicators too. Modern human women, for example, have breasts much larger than those of any other primate, yet their expanded size serves no apparent practical purpose. Female gorillas and chimpanzees have small breasts and nurse their offspring just fine. But for humans, full, round breasts subconsciously signal health and fertility. (The original meaning of the word *buxom* was healthy and easygoing, not large breasted.)[2]

The same is true for rounder rumps and a clearly defined hourglass figure. Several studies have revealed that men of nearly every culture are attracted to women whose waists are about 70 percent of the size of their hips. Other studies have shown that a certain amount of fat on the backsides and hips of women is a universal signal of fertility. Because of the subtle messages they send, over time evolution favored women with these traits for the simple reason that they were terrifically accurate indicators of health. Their offspring then tended to survive and go on to mate and pass the health-enhancing genes along.

Women seek out fitness indicators in men, too. They find slim hips and broad shoulders attractive because physical strength sends subliminal messages that such a man is not only a fertile source of first-rate DNA, but athletic enough to survive the dangers of the world and bring home the bacon.

The importance of fitness indicators has even driven the way we look. The human face is one of the best advertisers of health in nature, which is why we are so tuned in to beauty and handsomeness. We love symmetry in the countenances of others, not to mention bright smiles, white teeth, smooth skin, and thick hair. Many of us assume that we develop our attraction to these traits because we learn it. That's partly true. Fashion trends and hairstyles can affect what we consider beautiful as Darwin's research illustrated, but our tastes in physical beauty are almost entirely primal and subconscious, which is to say, they are not learned.

Psychologist Judith Langlois at the University of Texas, Austin, for example, has found these tendencies are so deep that even infants prefer comely caregivers to unlovely ones. She and her research team figured this out by gathering together the odd combination of sixty babies, one woman, and an expert mask-maker. For the experiment the team asked the mask-maker to fit two masks to a female caretaker—one that made her look pretty, the other not so pretty. This woman was a stranger to all the babies, and the masks were extremely realistic, a skin over the caregiver's real skin that smiled or frowned and moved seamlessly no matter how she expressed herself. To ensure that the woman didn't act differently depending on what sort of mask she was wearing—something that might subtly have affected the babies' behavior—the caregiver herself was never allowed to know whether the mask she was wearing made her appear handsome or unsightly. Only the babies knew.

Once she was properly disguised, the woman then began play with each of the sixty babies in turn. Their playing was tightly scripted to keep the experience for each child consistent. Every play date was captured on videotape, and low and behold, the study revealed that the infants, according to Langlois, "more frequently avoided the woman when she was unattractive than when she was attractive, and they showed more negative emotion and distress in the unattractive than in the attractive condition. Furthermore, boys (but not girls) approached the female stranger more often in the attractive rather than in the unattractive condition, perhaps foreshadowing the types of interactions that may later occur at parties and other social situations when the boys are older!"

Other studies reinforce the primal depth of our preference for beauty in one another. College students have been shown to prefer cuter babies to less cute ones, even when they initially said all babies look alike, and mothers have even been shown to act more attentively and affectionately toward firstborns who were considered attractive than to those who weren't. Grade-school children who are good-looking are treated better by their peers than their less attractive counterparts, and another Langlois experiment illustrated that babies no older than six months of age looked longer at pictures of attractive adults, no matter what their race or ethnic background.[3]

None of these experiments means that any of this behavior makes sense. In fact it is proof that it doesn't because the world is, regrettably, filled with attractive people who are neither kind, nor trustworthy, nor particularly intelligent, all useful traits in a human. Nothing about beauty makes it innately good or bad, and we have thankfully evolved the mental capacity to understand that. Nevertheless, we have a difficult time resisting the primal impulses that cause us to prefer physical attractiveness because it has proven over time to be a spectacularly strong indicator of a personal gene pool that endows its owner with a better chance of making it from one day to the next. It may not be as useful an indicator today as it once was, but millions of years of evolution creates habits that are wickedly difficult to shake.

Why does any of this matter? Because that childhood-extending phenomenon we call neoteny and our universal preference for beauty are profoundly bound to one another, even if it isn't immediately obvious. Together they help explain why the countenance you sleepily gaze at in the mirror each morning looks more like an infant ape's than a full-grown one's. Remember Konrad Lorenz's "innate releasing mechanism"? In addition to that observation, a surfeit of other studies reveal that infant faces, especially smiling ones, create a "pleasure response" in adults.

If that's true, then our more apelike ancestors may have begun to prefer mates who retained more youthful traits into their adulthood—higher foreheads, larger skulls and eyes, flatter faces, and stronger chins. Females that grew up by genetic happenstance looking more child-like would have found themselves with more enthusiastic suitors than other women who looked less childlike. That increased the chances that those baby-faced traits would be passed along to both female and male children, leading to still more neotenic looks in all of us.

But in addition to triggering caretaking and pleasure responses, youth is, as we know, also a fitness indicator. It goes hand in hand with health, strength, and fertility, giving members of the opposite sex still more reasons to prefer mates who retained their youthful looks beyond childhood. The process may have been long and slow, but over tens of millions of generations the simian appearance that had once defined our human ancestors morphed from sloped brows, protruding snouts, and receded chins into more childlike traits. We can see exactly this transformation in the faces of our ancestors, species by species, as we march from the deep past toward the present. By the time *Homo sapiens* had emerged two hundred thousand years ago, our youthful looks had pretty much reached their current state.

A Preference for Youth Is Still
Shaping Our Evolution

If more proof is needed of our preference for youth in potential mates, a study performed by scientists in Scotland, Japan, and South Africa seems to have supplied it. You may not find it terribly surprising that the research uncovered that men prefer women whose faces look more feminine, which is to say youthful; but it also turns out that that women preferred *men* whose faces looked more feminine, or boyish.

For the study, scientists digitally created an average, but attractive version of two faces for each sex, one Caucasian and one Asian, four "average" faces in all. They then digitally modified each face to create two versions, one slightly more masculine, the other slightly more feminine and childlike. The changes are subtle, but the male versions of the faces sport slightly heavier eyebrows, a hint of shaved beard, squarer jaws, and pupils that stand a bit farther apart than female pupils, something that tricks the eye into thinking that the male faces in the study were larger than their feminine counterpart (they weren't).

When forced to rate the faces they found most attractive, members of both sexes, old and young, Asian and Caucasian, said they preferred the more feminine versions. In addition, when prompted to rate the faces on something more than attractiveness, such as trustworthiness, warmth, cooperativeness, and the likelihood to be a good parent, again the more feminine faces were preferred, although youthful looks didn't seem to make the study's participants feel that feminine faces were either more or less intelligent.

If we have such a universal preference for feminine, youthful looks, then why don't men and women today, after millions of years of evolution, look essentially identical? Because some other factors are involved. A man's bigger body, larger muscles, and broader shoulders can also indicate a good protector and provider. Those traits require more testosterone, and more testosterone causes changes in a man's face you don't see in a woman's; a beard, thicker eyebrows, broader jaws, and a bigger head, for example. So, while male and female *Homo sapiens* look more like one another than any other humans, and certainly more than full-grown great apes, we don't look identical. But in time we might because, clearly, even today, we remain genetically predisposed to find younger, more childlike faces attractive.

The slow realignment of our looks over millions of years may have caused us to appear more childlike, but as we evolved and became more self-aware, apparently even this failed to make us attractive enough, because for at least the past fifty thousand years we have creatively and enthusiastically taken matters into our own hands, modifying our looks without waiting for genes and evolution to get around to the job. You and I can't take much credit for the blue eyes or blond hair, long, thin bodies or the round, stout ones, that our parents passed along to us for the simple reason they are nothing more than a genetic toss of the dice. But the extravagant measures we take to enhance our appearance that Darwin studied so exhaustively illustrate, pretty dramatically sometimes, something that we do that other animals don't, even other primates. We imaginatively elaborate our appearance, which may be one of the key behaviors that separate us from the rest of the animal world. But what is even more intriguing is that we don't simply tinker with our looks; we change our behaviors, too. We don't simply try to *look* sexy, we try to *act* sexy, and that, as a species, has taken us into entirely new territory.

Some of these modifications have obvious animal analogues, but with distinctly human twists. For example, why roar like a lion when you can show up on a date with a Porsche Carrera, or Harley-Davidson Night Rod Special. We not only use clothing to protect us from the elements, but also to improve our looks and make statements about status, power, and confidence. These behavioral elaborations even help to explain our affection for what sociologist and economist Thorstein Veblen termed "conspicuous consumption" in his landmark 1899 book *The Theory of the Leisure Class*. Possessions—the latest smartphone, the most fashionable house, the biggest diamond, the hottest dress, the richest fabrics—are all human made fitness indicators.

You might say, well, this all seems fairly banal, finding ways to make yourself more attractive to the opposite sex, and I am willing to agree. But you can make a powerful argument that efforts to enhance and amplify ourselves to impress potential sexual partners laid the foundation for far more creative endeavors, undertakings that have made much of modern human culture possible—song, art, invention, wit, storytelling, and humor. It could be argued that the foundations of human creativity and culture can trace their roots to our early efforts to consciously make ourselves more irresistible.

Psychologist Geoffrey Miller has argued that just as shapely bodies and symmetrical, youthful faces signal physical fitness, creativity itself is a sign of a mental fitness, something that has enormous value to a potential mate, and therefore a trait that evolution would "encourage."

Of course the organ that is the engine of all of this creativity is your brain. It may have evolved to make sense of the world you live in, but among us *Homo sapiens* it has become extraordinarily effective in generating all sorts of appealing behaviors and countless personal decisions that make you cooler, sexier, and downright captivating. It enables you to be witty, conjure startling ideas, master the piano, helps you dance better or sing beautifully, or become a more stable and loyal partner. It's a kind of universal machine that can turn itself to nearly any goal, including the sexual capitulation of the opposite sex. Once our brains found themselves self-aware, Miller argues, it emerged as nature's ultimate indicator of fitness. Just as genes can deliver vibrant feathers or neon colors, brains bent themselves to the work of upping our desirability in a million different ways. He calls this the "healthy brain theory."

The idea that creative behavior makes us sexier isn't brand-new. The old master Darwin, keeping in mind the antics of prancing and warbling birds, speculated in *The Descent of Man* that humans used both dance and song to win the hearts of potential mates. "I conclude that musical notes and rhythm were first acquired by the male and female progenitors of mankind for the sake of charming the opposite sex. Thus musical tones became associated with some of the strongest passions an animal is capable of feeling . . . We can thus understand how it is that music, dancing, song and poetry are very ancient arts."[8] In another part of the book he writes, "As neither the enjoyment nor the capacity of producing musical notes are faculties of the least use to man in reference to his daily habits of life, they must be ranked among the most mysterious with which he is endowed . . . Whether or not the

half-human progenitors of man possessed . . . the capacity of producing, and therefore no doubt appreciating, musical notes, we know
that man possessed these faculties at a very remote period."[5]

In other words, while there don't seem to be many practical reasons
why talents like music, dancing, and other arts evolved, still they did,
so there must have been some powerful selective forces at work to
bring them into existence, winning over the opposite sex, for example.

Zoologists have an oddly charming name they use to describe the
singing or dancing or fighting that animals do to gain the attention of
potential mates. They call it lekking. It's a way of strutting your stuff,
letting the creatures you are wooing—not to mention any competitors that happen to be nearby—know just how fit and cool you are.
When we stand around at a party and talk, the human version of lekking is rampant and intricate. We show off the way we look and
dress, revealing subtly (or not) the clothing or jewelry we wear or the
gadgets we have on hand. But the real action is in how we behave.
Are we funny, insightful, charming, articulate, and quick-witted? If
we are, we are advertising a first-rate mind. The more talent and creativity we bring to the party, the more likely we are to be noticed.
Being outstanding is a good thing when vying for the attention of
others.

We cultivate these behaviors in subtle and complex ways that even
we aren't consciously aware of. Researchers have found that women,
for example, laugh more when they are in the company of men. This
isn't because men are exceptionally funny, but because (subconsciously)
women are encouraging men to lek so they can gather information and
observe what the man has to offer. The more she laughs, the more he
shares and reveals. And the more he reveals, the better she can judge
what he offers in ideas, values, talent, and personality. If she likes what
she sees, she may eventually offer him the benefit of her company. If
not, the laughter stops and she moves on. This probably also explains
the results of a 2005 study that indicated that women are attracted to
men who make them laugh while men are attracted to women who
find their jokes funny.

A recent study of 425 British men and women indicated that artists,
poets, and other creative "types" had two to three more sexual partners than the average Brit who participated in the study. Whatever
else you might conclude about bohemian lifestyles, it seems that cre-

ativity has its attractions. Another study has found that professional dancers (and their parents) share two specific genes associated with a predisposition for being good social communicators. The theory here is that dance and song were primal ways that our ancestors bonded, prepared for battle, or celebrated, and that creative dancers not only boasted great rhythm, but great social skills, which together made them especially attractive. This would make dancing both a way to show off physical fitness *and* a healthy brain, a kind of evolutionary twofer. Could it be that charm, creativity, and rhythm all go hand in hand?

We can speculate, but the truth is it has been a struggle for scientists to take behaviors such as art, sculpture, storytelling, and music seriously because each seems, from an evolutionary point of view, so impractical. They also resist cold analysis because they are hopelessly subjective. Mostly the field of evolutionary psychology has concluded that music, song, dance, and art are best explained as accidental by-products of other forces that created the extravagant human brain. Nothing more than evolutionary filigree.

But again, Geoffrey Miller begs to differ. He argues that our elaborate human behaviors evolved for the same reasons peacock feathers did, or the rainbow colors on mandrill snouts—they represent powerful personal marketing that lets the opposite sex know how extraordinarily fit the brains of their owners are, which in turn makes them great potential mates. "The healthy brain theory," he says, "proposes that our minds are clusters of fitness indicators: persuasive salesmen like art, music, and humor, that do their best work in courtship, where the most important deals are made."

I believe that Miller is correct, but I also believe that advertising our cerebral fitness is good for more than landing mates, as crucially important as that is. In fact, creativity of all kinds may trump sex as *the* most central force in human relationships because, beyond sex and sexual selection, survival is also, ultimately, about power over your environment. And fit brains not only demonstrate power, they generate it.

In 1975, Amotz Zahavi, a biologist at Tel Aviv University, conceived a theory that was fascinating because on the surface it was so counterintuitive. He thought it might explain some of the exceedingly impractical traits and behaviors we see in nature that seem to hamper animals rather than help them. Why, he asked, would peacock feathers evolve when they weigh so much and their colors risk attracting the attention of predators? Or why, when an impala senses a lion nearby, does it bound straight up in the air (something called stotting), wasting valuable seconds before it sprints in the opposite direction? Why do bowerbirds create intricate and ostentatious nests for their mates that include everything from seashells to rifle shells when a simple bundle of woven grass would do the job just as well? To answer these questions he conceived the "handicap principle."

Zahavi already knew some of the traits and behaviors could be explained as ways to win mates. But he also knew they help establish status. The peacock isn't simply saying, "See my remarkable feathers." He's also saying, "And have you noticed how strong I must be to get off the ground and fly with these enormous things weighing me down?" The point for potential mates is clear—I'm handsome *and* strong. But the same message is simultaneously sent to predators and other peacock competitors: "Don't mess with me. I'm top dog. I know it. You know it. So let's all just take our place in the pecking order and move on."

In the same way, an impala's pogo-stick bound before it sets out to escape from a predator may waste time and energy, but it also tells a stalking lion, "As you can see, I'm pretty healthy and rather quick. You might want to think twice before taking the time to chase me." Often as not, the lion does a quick and primal cost-benefit analysis, walks away, and looks for a less challenging meal elsewhere. These are survival strategies, pure and simple.

The point is, even seemingly inefficient traits and behaviors have their purposes, though they might not be immediately obvious. It's not always necessarily about sexual selection. Sometimes the traits make you attractive, sometimes they represent a sophisticated way to survive, sometimes they help reinforce status, and sometimes it's all the above.

If ever there was an example of an organ that was costly, yet delivered an enormous payoff, the human brain is it—the ultimate peacock's feathers. It devours enormous amounts of energy (far more than any other organ in the body), is outrageously complex, and subject to breaking down (with disastrous results). Yet what powerful messages it can send about its owner, and its owner's fitness! This makes the human brain the most elaborate example of the handicap principle in all of nature, an extravagance that expends enormous amounts of energy illustrating how extraordinary its owner is by conjuring the most surprising and creative things it can itself conceive.

How else can you explain Beethoven's Ninth, Picasso's *Guernica*, and sculptures from Michelangelo's *Moses* to the great and intricate Buddha of Kamakura, Japan. Why Fred Astaire, Kabuki Theater, James Joyce, Cirque du Soleil, Steve Jobs, Gregorian chant, and *Avatar*? In short, how do you explain all the seemingly impractical yet ubiquitous examples of human creativity and inventiveness?

Because the brain is invisible, unlike peacock feathers, it reveals its fitness by generating behaviors that are extra-ordinary, surprising, and impressive. To be surprising means to be different and unexpected, again, out-standing. To be impressive the behaviors have to be something others find difficult to do. The two together define creativity. The scale of human invention is broad and deep. It can encompass everything from the merely pleasing to stunning genius.

When you think about it, the brain's capacity for generating captivating insights and behaviors is what makes each of us the unique people we are. We use it to fabricate the traits that define us—our wit, our charm, our drive, our insight, our humor and intelligence, our talent and interests. Some of us have been blessed with truly extraordinary gifts—Shakespeare, the ultimate storyteller; Leonardo, the ultimate imagineer; Einstein, the ultimate problem solver. The rest of us stake our ground somewhere between profound genius and a good one-liner.

Why is this need and appreciation for creativity so deeply plaited into us? Because the advantage of a brain that can do surprising, remarkable, or outrageously pleasing things is that it gets attention, or rather its owner does, and that attention can be translated into fame, influence, goodwill, leadership, sex, and, in modern society, money. Look at the people we admire or reward across all cultures. Dancers, singers, thinkers, comedians, actors, political leaders, entrepreneurs, and businesspeople, even an occasional scientist or journalist. (I am not including athletes here because we don't reward them for their intelligence, though their intelligence may certainly contribute to their success.) All of these people display unusually fit brains because they are both inventive and able to effectively communicate their inventiveness. Whatever else we may think of them, we have to at least agree that they are not boring and or predictable. They stand out, and in standing out, they aggregate the most important human commodity of all—power.

We often think of power as a bad thing, possibly because it can be abused with depressing effect. But in nature acquiring power is crucial to survival. All living things seek it because without it they will die. Plants may acquire it in the form of nutrients from the soil and the sun. A silverback gorilla or bighorn sheep may acquire it with raw strength. With most animals power flows to them in direct proportion to how well the genes they inherited match their environment. Penguins would be powerless in the tropics, and Komodo dragons would be equally helpless in the arctic. Cheetahs maintain power with speed, wildebeests in numbers, and condors with flight.

But we humans apply our brains, not simply our genes, to acquiring power, and because we are so genial, we seek it not only to survive our physical environment, but our social one, too. Survival in a social context isn't quite as literal as it is in a physical one. If you don't survive physically, you die. If you don't survive socially, it means you don't matter, and that is, in it's own way, also deadly.

Mattering is itself relative because in today's world we can live in a wide variety of social circles. We can't all matter as much as those examples I mentioned earlier, Aristotle or Confucius or Einstein, Leonardo da Vinci and Shakespeare—people whose creativity made an indelible mark on human history. But we can matter to our city or officemates or family or Facebook friends—the modern equivalents of the tribe—and that is important because how we stand with our tribe deeply affects how we feel about ourselves. Today we can even have multiple tribes to choose from, and the World Wide Web allows us to create instant new tribes to whom we can display our cerebral fitness. The important thing is that we matter, to someone. Because if we don't, the alternative is chronic depression, or worse.

Creativity isn't the only way we strive to matter and gain power, but it's the most functional, sensible way. It doesn't require greed or jealousy, envy or outright violence, all of which can be highly effective, if immensely damaging, methods for gaining power. But these don't reveal a fit brain. Creativity does. It is the most impressive way to earn the attention of others. And thankfully, over the long haul, it works; otherwise it would long ago have been swept from the index of our behaviors. We would be without art, music, and dance; there would be no pyramids at Giza, no Taj Mahal, no Brahms, Voltaire, Goethe, Yeats; only brutality and violence, and therefore, very likely, no humans.

The idea that the foundations of human civilization are largely an unintended consequence of complex brains wired to draw attention to their owners is both paradoxical and startling. Brains did not evolve to be creative, they are creative by the accident of evolution. And in becoming so, the exciting and innovative sideshow that bubbled up from our primal need to matter to the opposite sex, our competitors, loved ones, and everyone else in our tribe eventually took center stage. Now, after thousands of years of our brains' showing off, we find ourselves enmeshed in this massively complicated, rich, and remarkable thing called human culture, sometimes revealing the evil in us, sometimes the divine, but always surprising and innovated because we have become utterly incapable of living without originality. There is no getting around the conclusion that creativity, though it may once have been evolutionary filigree, has become *the* force that defines our species, and the behavior that separates us from all other living things.

As creative as we are, we haven't yet solved the elusive question of when, or how, we managed to get this way. It's not as though evolution one day snapped its fingers and we were smitten. The cerebral infrastructure that makes such a thing possible has been long in the making. Nevertheless, evidence of human creativity in the sense we are talking about has been scarce until quite recently, if you can consider recently within the past seventy thousand years. It's true tools and other technologies had been around millions of years, and they require creativity, but they are not examples of self-expression or symbolic thinking the way a piece of sculpture, a painting, language, or a song are. The timing of this matters because creative self-expression of this kind only became possible when our brains reached a certain critical, but as yet undefined, level. Its emergence marks a watershed event in human evolution, arguably *the* watershed event.

Most paleoanthropologists agree, for the time being at least, that *Homo sapiens* emerged 195,000 years ago. By this they mean creatures that were anatomically modern—they looked like us. The oldest *Homo sapiens* fossils were found in Ethiopia in 1961, but sadly no trace of symbolic thinking was found with them, no tangible demonstrations of brain fitness. This has created the underlying suspicion among scientists that though these people looked like us, they may not have *acted* altogether like us. They made tools that were incrementally better than the tools of those who came before them. They certainly lived rich and complicated social lives. But all the fossil and genetic evidence indicates that mostly they still roamed the same grasslands in East Africa, hunting game and struggling to survive, as so many of their ancestors before them.

For over a hundred thousand years the first of our kind lived this way, resembling you and me physically, and perhaps in many ways emotionally, but apparently not mentally. It was as if the brain had reached regulation size, but hadn't yet completed all the wiring and biological alchemy needed to summon up a mind that saw the world quite the way we do. This has been a gnarly problem for scientists because you cannot fathom the minds of creatures with whom you haven't the luxury of sitting down and talking.

Around seventy-two thousand years ago, on December 27 in the Human Evolutionary Calendar, we begin to see the evidence of a change in what might have been a hotbed of rapid human, intellectual development—those coastal cave communities of South Africa where, according to Curtis Marean, small *Homo sapiens* communities found themselves within a gnat's eyelash of total annihilation.

At Blombos Cave the evidence tells us that a small handful of *Homo sapiens* were decorating tiny nodules of hematite, a kind of iron rock, with geometric designs, cross-hatchings that may have represented some kind of symbol, still indecipherable to us. In the same cave, but later in time, scientists have also unearthed perforated ornamental shell beads, arguably the first evidence of human-made jewelry. These discoveries were made in the 1990s and early 2000s, but then in 2010 a team of paleoanthropologists reported finding nearly three hundred fragments of decorated ostrich eggs in the Diepkloof Rock Shelter, another South African cave complex. Each shell is sixty thousand years old, and each was painstakingly etched with precise crosshatched designs, proof, the team believes, that the people who made the markings considered them important symbols. If this theory is correct, the cross-hatchings found on rocks twelve thousand years older may be more than meaningless doodles, as some scientists suspected when they were first found. Did they contain some secret message? Words, perhaps? Or calculations? An early form of sheet music, maybe? Or someone's grocery tab? Their significance remains elusive, but enticing.

Despite these clues, and some scattered signs that Neanderthals in Europe had attained a semblance of symbolic thinking, the evidence for creativity of the indisputably modern human variety doesn't begin to appear until around forty thousand years ago, and by then the evidence is both stunning and global. By this time *Homo sapiens* had made their way out of Africa for good and were busily populating Europe, east and south Asia, and making their way through Indonesia clear to northern Australia. There on the rock walls of Australian caves, ancient humans began to paint symbolic figures and animals, having improved, perhaps, on the creative habits of their ancestors from Africa who had found and used ocher or cryptically symbolized their feelings and insights on the shells of ostrich eggs.

Afterward more proof of symbolic thinking begins to surface. Archaeologists have found small but remarkable sculptures, sometimes of penises, but more often of large-breasted, pregnant women carved by talented artists, beginning thirty-five thousand years ago. They call these Venus figurines because they seem to be talismans of fertility, a trait undeniably crucial to a species who certainly found strength in numbers, but whose life spans rarely reached beyond their thirties. Most of the objects are small and portable, custom-made, perhaps, for magically connecting with the mysterious forces of nature. From Western Europe to Siberia anthropologists have found these small sculptures, and along with them figurines of chimeras—half-human, half-animal—all astonishing indications of a mind unlike anything the powers of life had produced in the long course of their 3.8 billion years of existence. Creatures that could not only imagine other worlds, beings, and forces, but express their imaginings, in the hope, somehow, that they could tap the strength of those mysteries.

Some of the most breathtaking art was created by the Leonardos and Michelangelos of their time deep in caves in Lascaux and Chauvet, France, and Altamira, Spain, as the last great ice age began to release its frigid grip on Europe. These images would be the envy of art galleries around the world today, or Madison Avenue marketeers—rich, vibrant, and ingenious. You can almost see them move and ripple in the flickering firelight that once illuminated the cave walls as the

Cro-Magnon artists stood with their palettes of primordial paints and dyes, dabbing the walls, extracting the beasts from their minds and applying their images to the rock. What powerful magic this must have been to the painter and those who witnessed the work. How could any creature imagine such things and then make them appear right before your very eyes? What hidden powers could enable a living thing to consciously and purposefully create beauty out of nothing more than the popping of the synapses in his head?

So far more than 150 caves have been found in Western Europe, primal cathedrals where the walls have been saturated with the conjurings of artist humans showing off the startling fitness of their brains. We can only imagine how revered people like these must have been, made powerful because from their fingers flowed the symbols of the beasts that fed and clothed and killed these itinerant hunters. How out-standing they must have seemed.

The purpose of these paintings remains a mystery. Colored footprints of both children and adults that show up on the floors of some caves signify, for some, that rites of passage were performed here as boys made their transition to manhood, or girls became capable of bearing children. Others seem to have been a kind of play school for ancient human children dabbling in the art of art.

Some have wondered if the images became a way to control the creatures they depicted, or to draw out and drink in their predatory strength. Maybe these were the theaters of their day, where great stories were told of heroes and their exploits, or a place where men hunted, virtually, in a kind of primeval video game, imagining with their paintings the ways they would bring down prey when, at last, the long and punishing winters ended. Strangely, the cave paintings almost never depict a human form, and when they do, the figures are sticklike, as if humans are minor players in a larger drama. Are these the remnants of a creeping epidemic of human creativity, isolated breakouts of beauty? Are they examples of a new kind of mind, self-aware, curious, and brimming with ideas and emotions, that had no choice but to express itself for the pure joy of it, like a child playing with crayons, or a graffiti artist saying, "I'm here! And I matter."

These settings, perhaps because they are encased in rock and filled with the ghostly work of their artists, feel sacred and magical. It's easy to imagine ceremonies of some kind taking place within the bowels of the earth accompanied by chants and primeval music. Archaeologists have found drumsticks, flutes, and a prehistoric instrument called a bull-roarer near the caves of Lascaux. You can hear the rocky acoustics amplifying the chants and music, the drumsticks beating out a steady rhythm accompanied by the eerie thrumming of the bull-roarer, a sound like the breathing of some great sleeping beast, all combining to make a powerful and ancient symphony that moved and bonded the new kind of primates who listened.

Music may be the most ancient of human arts. Chanting and dancing were arguably practiced by tribes of *Homo erectus* over a million and a half years ago, and later by *Homo heidelbergensis*, the common ancestor of both *Homo sapiens* and Neanderthals, seven hundred thousand years ago. Thirty-five thousand years in the past, dancing and music had likely become much more complex than the varieties our more ancient predecessors practiced, a way to entertain and express personal feelings as well as to bond and celebrate.

The importance of dance and music in the human psyche is probably best illustrated by a single startling fact. We are the only primates that can tap our foot or move our body in time with a specific rhythm. It's wired into us, but not into our chimp or gorilla cousins, which tells us that it is a trait that like language, big toes, and toolmaking evolved sometime over the past seven million years.

It's difficult to explain why *Homo sapiens* took more than a hundred millennia to show off the creativity that stands as the irrefutable proof that the stock from which you and I sprang had truly arrived, but that hasn't stopped it from being passionately debated. Some paleoanthropologists argue that an explosion in *Homo sapiens* population seventy thousand years ago eventually generated competition that in turn encouraged innovation. Others believe that there was no "big bang," no sudden blossoming of human creativity and symbolic thinking at all. Instead we are simply seeing the slow and aggregated results of gradual human progress that finally left behind enough proof in the fossil record that it existed. Others have argued that as the human race grew, creative ideas that had once been conceived but later lost were now picked up and passed along more easily. More of us were around to ensure that great ideas were absorbed, reused, and built upon rather than wiped out when the innovator passed away.

Another possibility exists. Stanford paleoanthropologist Richard

Klein holds that the catalyst for human creativity didn't happen out-side in the real world, but inside our heads—a genetic mutation, or series of them, that transformed the way our brains functioned so that symbolic thought and the creativity it makes possible erupted from our ancestors' minds like Athena from the head of Zeus. Somewhere, somehow, he believes, the wiring or the chemistry of the brain changed, perhaps subtly, and crossed an invisible threshold that made it possible for us to attach complex meaning to otherwise meaningless pictures, objects, or sounds. Images could represent gods; beads and shells could represent value; shapes could stand in for ideas that anyone who saw them would mutually, and immediately, understand. Sounds could become symbols for words, and symbols could be built into the grammar and syntax that make language the remarkable thing it is.

Once this happened, says Klein, "humanity was transformed from a relatively rare and insignificant large mammal to something like a geologic force." The mechanisms for this change are unknown. It could be random genetic mutation, or, as University of Cape Town archaeologist John Parkington theorizes, a new kind of diet. Parking-ton believes it is not a coincidence that the early humans in South Africa who were making jewelry from seashells were also eating large amounts of seafood out of those very shells, and that food was provid-ing the fatty acids that we today know are crucial to brain health and function. The new sources of food, he believes, combined with a more modern cerebral architecture than earlier humans, made these *Homo sapiens* "cognitively aware, faster-wired, faster-brained, smarter," and their seashell jewelry, art, and technical advances stand as the proof.[6]

There is evidence that the chemistry of the modern human brain, especially the prefrontal cortex, the most recently evolved part of us, operates differently from that of other primates. When scientists in Shanghai, China, compared one hundred chemicals in the brains of humans, chimpanzees, and rhesus macaques, they found that the lev-els of twenty-four of them were drastically higher in the human pre-frontal cortex. It would be interesting to know how these levels would compare to those in the brains of Neanderthals, *Homo ergaster*, or even *Homo sapiens* who lived more than seventy-five thousand years ago, but, of course, none of those specimens exist. Would we find that somehow the brain had leaped chemically forward, allow-ing us to cross some unknown hormonal Rubicon? The findings indicate that when it comes to glutamate, the main excitatory neurotransmitter in our brain, we modern humans are in a league all of our own, constantly burning vast reservoirs of it compared with other primates. This may reinforce Parkington's theory that something has made us "faster-brained."

As it happens our penchant for inventiveness is also linked to our species-wide predilection for youthfulness. That shouldn't surprise us. When you look at creativity in action, it bears a close resemblance to a child at play. One of its hallmarks is that concepts, thoughts, words, or objects that don't normally go together are joined in novel ways and result in something that is useful or arresting or jaw-droppingly beautiful. When these coalitions come together in a eureka! moment, something that once seemed improbable now stands, right there, real and complete.

For children nearly everything in the world is new, and so almost any combination of unfamiliar experiences can result in those moments of discovery. Since so much is unfamiliar in a child's experience there is enormous room for learning. But as we grow older and experience more, the space for true innovation narrows, and the stakes rise. The creative bar becomes trickier to reach. Startling is tougher to come across. Still, we humans manage to do it every day, day after day. And the reason we do is because, of all the apes, we are the most childlike.

By shifting the time when genes express themselves, and by rearranging brain and hormonal chemistry, neoteny not only transformed the way we look, but the way we act. Cognitive scientist Elizabeth Bates wrote about the power of neoteny and its ability to generate powerful change in 1979, but at the time she didn't connect it with creativity; she associated it with another benchmark event in human evolution, language. She (and others) believe that a human "language acquisition device" evolved, like nearly everything else in life, by recombining a variety of preexisting capacities into a new configuration. Human language, she argued, was built on the shoulders of "various cognitive and social components that evolved initially in the service of completely different functions . . . [and] that at some point in history, these 'old parts' reached a new quantitative level that permitted qualitatively new interactions, including the emergence of symbols." Put another way, neoteny helped shift the growth patterns of one or more capacities our ancestors already possessed for interacting with one another and commandeered them for new uses.[7]

If neoteny played a central role in the emergence of language, could it also have played an earlier role in the ingenuity that symbolic thought requires? It's possible. The timing of the expression of certain genes, including genes that control brain growth, made and makes our long childhoods. It extends the time our brains are pliable and able to bend to our personal experience. But because human neoteny is *so* extreme, it has done even more than that. While it acts most powerfully during our childhood and makes childhood possible, it also extends childlike behavior throughout the long course of our lives. Even in old age, we are more childlike than other primates are in their youth. The brain flexes and muses and creates right up until the end. "We don't stop playing because we grow old," the aging playwright George Bernard Shaw once mused, "we grow old because we stop playing."

This means we are not only children longer, we are childlike longer, and that has made us by far the most creative and adaptable creatures ever. "We are not a computer that follows routines laid down at birth," Jacob Bronowski once observed. "If we are any kind of machine, then we are a learning machine."

This is why child's play and creativity are so deeply linked. *Play* has multiple meanings depending on whether you are an anthropologist, psychologist, parent, or child, but among its hallmarks are the simple joys of pushing boundaries, expanding limits, randomly galumphing around to see what happens just for kicks. Even long-faced philosopher Martin Buber had to admit, "Play is the exultation of the possible."

At the heart of playing is the strange phenomenon of curiosity. You really can't have one without the other. One theory about curiosity is that we are all born "infovores," that we crave new knowledge and experience in something like the way we crave food. It's a kind of mental and emotional hunger that requires ongoing feeding and satisfaction. Old knowledge doesn't satisfy our curiosity because it's familiar; we have "eaten" it before. So how do we know when something is new? Because it surprises us, because it's different from what we are used to, fresh.

Every creature has an evolved talent for identifying what is surprising or out of the ordinary for one simple reason: it's central to survival. Those that fail to tune in to the change around them, those that aren't sensitive to surprise, soon join the legions of species no longer with us. It's a talent that reaches back hundreds of millions of years.

For modern humans like you and me this makes curiosity a way to gather new information that has survival benefits, but also a process for gathering the building blocks out of which we assemble entirely new experiences and new forms of knowledge. One of the behaviors that makes us different is our affection for playing around randomly, joining this with that or that with another thing with no particular reason except to create more surprises that satisfy our curiosity, which in turn results in still newer experiences, new inventions and insights. Innovation and originality are by-products of our lifelong, childlike love off goofing off!

In some ways, play resembles evolution itself, randomly introducing unpredicted and unpredictable innovations the way random mutation reshapes DNA. When you think about it, adaptation in nature is a kind of learning. Something different comes into the world, and living things adjust genetically. The adjustment is serendipitous, not conscious, but it happens.

Play does something similar. It randomly introduces new experiences to our minds, again and again. We encounter novelty, and when we find it useful or enticing, we make it ours. It literally changes our minds, and therefore us. And since not one of us learns quite the same things, since each of us plays in different ways and is surprised by different experiences, your changes of mind are different from mine, which makes each of us unique. Our view of the world is not entirely distinct, but distinct enough that we ourselves become new and surprising additions to it. This also means that you and I can learn from one another by sharing our differences, a little like the way two parents' different chromosomes combine to create a genetically unique child. By acquiring new experiences and then sharing them, ideas and originality become sticky and spread from mind to mind.

No matter how long we live, we can't seem to root the child out of us entirely, joyful in its experimentation, never satisfied, hungry for knowledge, and eager to show it off. When you look at us this way—a lifelong child, with a mind itching to play, and famished for surprise—you can see how the power for creating originality out of random experience, and the ability to share those experiences, could have taken us from a mere ten thousand or so primates seventy-five thousand years ago scrambling back from the abyss of extinction, to seven billion creatures who have not only populated every corner of the planet, but managed to rocket away from it a few times to orbit and land elsewhere in the solar system. By connecting the surprising experiences and ideas we spawn or stumble across, and then sharing them with one another, we have been able to construct great edifices of new knowledge—Pythagoras's geometry, Newton's and Leibniz's calculus, the wheel, clocks and longbows, the Saturn V rocket and the silicon chip and balalaikas, silk paintings, the telescope, money, sailing ships and steam engines, kissing and language, music of all kinds and toys of every imaginable stripe, chess, baseball, sculpture, and van Gogh's *Starry Night*—all of it out of the combined, interlocked, unique imaginings of millions of minds shaped by billions of surprises shared in trillions of exchanges to create the chaotic, astonishing, tumultuous stew we call human culture. In this sense, we are a race of continually startled, and startling, creatures.

Once the adaptable nature of such a pliable human brain had been sufficiently honed and wired to make all the improbable internal links needed to connect "new" into still newer creative acts, human culture was guaranteed to evolve at exponential speed.[8]

However it all happened exactly, clearly something radically different was emerging in the brains and minds of *Homo sapiens* from Europe to Africa to Australia between seventy-five thousand and forty-five thousand years ago. Some sort of cerebral critical mass was frothing. Neoteny had created a nimble, pliant brain that remained flexible throughout life and generated both unique people and unique ideas. We had evolved into born learners, genetically encouraged to seek out and, by some strange neuronal alchemy, devour surprise and transform it into knowledge.

This may be why we, and not Neanderthals, are still around today to wonder where we came from. It may explain why you are at this moment gazing at a page of symbols I have typed that your mind, rather astonishingly and without much seeming effort, translates into thoughts you can understand.

Neanderthals lived faster than we did and they died younger, and possibly therein lies the reason we remain and they don't. Though they, too, were neotenic and time had also been genetically rearranged for them so that they were born earlier and remained young longer than today's chimpanzees, gorillas, and orangutans, their childhoods were not as long as ours. This gave their brains less time to shape their personal experience, their ideas, and their personalities before they began to grow more rigid. And growing more rigid, they may have been a less childlike species, less prone to experiment. That would have made them less adaptable. Perhaps this was also true of the Denisovans, and the Red Deer Cave people of south China, even the "hobbits" of Indonesia. Their minds may have been as sharp, but not as plastic, as those of the *Homo sapiens* who had recently migrated out of Africa. Perhaps they all became more set in their ways sooner; more adult, you might say.

Neanderthal tools, and the little we know of their rituals, indicate they were on the cusp of our brand of symbolic thought, but some pieces, we don't know how many, didn't quite fall into place in a way that allowed them to remain among us today. If their language was songlike as Mithen has theorized, they might have expressed the emotions they were feeling, more than the explanations of why they were feeling them. There might have been more passion, less logic, or maybe less of a balance between the two.

We can imagine them as a bright, lyric, almost mystical species, but not a fully symbolic one. Perhaps they lived in a kind of surreal, Daliesque world, less self-aware, not altogether capable of encapsulating the ephemera of the new thoughts their minds conjured into carvings, sculpture, patterns, or images made from strokes of paint. Because that is what symbols do, they translate thoughts and ideas into tight, little packages of meaning for delivery from one mind to another. It is miraculous really.

Maybe the Neanderthals weren't radically different from us, or less intelligent; they may simply not have been able to play their way into symbol making as complex as the kind we stumbled upon. In particular, perhaps they couldn't play their way into the most shattering gift of all—spoken language as we know it today, complete with the bells and whistles of grammar and syntax. Perhaps.

Our youthfulness, our propensity for playing with, and juggling and shuffling, surprising experiences and insights continually and in more startling incarnations must have cried out for an invention as elegant as language. When you step back from it, language is something like a piano. Using nothing more than a piano's eighty-eight keys, a player can express an infinite number of songs, and infinite variations on those songs. With language we can express an infinite variety of thoughts, feelings, ideas, and insights. Before modern language, our ancestors may have been capable of gesture, art, and song with which to bundle and share the flickerings of their minds, but imagine, how modern language must have supercharged human creativity and the culture that was assembled out of it?

The thing is, while language connected us to one another more closely than ever, it also enabled us to pull off another remarkable feat: it made us aware that we are aware. It may also have made madness possible.[9]

Chapter Eight

The Voice Inside Your Head

I am a strange loop.
—Douglas Hofstadter

IF YOU COULD shrink down to the size of a molecule and slip into your brain, you would find yourself flying among billions of neurons along great highways of dendrites and axons with streams of chemicals splashing across synaptic gaps and firestorms of electricity arcing all around. At this scale, the real estate of your mind would be vast, planetary in its dimensions, as you rode your molecule-size vehicle. Everywhere commands that make it possible for you to walk, breathe, see, smell, speak, reflect, and imagine would be at work.

Witnessing the weather of your thoughts and feelings like this would be extraordinary, but even from this vantage point, or maybe because of it, you could never imagine that all of the impulses and chemistry blowing up and down the intricate infrastructure around you could possibly *be* you. Yet it is. You are assembled from these nonstop, chaotic processes; the rolled-up, aggregated chemistry and biology through which you are zipping. Stupefying but true.

The bewildering mystery of how this happens is what Douglas Hofstadter hoped to resolve when he penned his landmark book *Gödel, Escher, Bach*. He wanted to figure out, he wrote, "What is a self, and how can a self come out of stuff that is as selfless as a stone or a puddle? What is an 'I'?" It's an easy question to ask, and we've been asking it for as long as we have been around. The answer, though, is just a bit tougher to come by than the asking.

But we can try.

You may have noticed from time to time that when you think, you find that you are talking to yourself; not necessarily out loud, but in your mind. Nearly every waking moment we describe what is going on in our minds to ourselves, like a sports announcer calling a game, remarking on what we see, commenting on our own insights, planning our lives, polling ourselves on what we feel, wondering why this and how that? "God, I'm edgy this morning. This coffee is delicious! Hmmm, rain, better grab the umbrella. That's an interesting choice in a hat, if you are insane! Don't forget to get the oil changed and pick up the milk. You know you really have to get better at remembering people's names." Our reflections run the gamut from the mundane to the ethereal, occasionally the sublime, but they almost never stop, until we fall asleep.

When you think about how you think about yourself, you are experiencing what psychologists call metaconsciousness, the ability to be aware that you are aware. Though we take it for granted, this capability requires language, and the interesting thing about that is that language requires using symbols that we sound out in our minds so we can understand what we are saying to ourselves. That we can do this is remarkable enough, but doesn't conversing as we do with ourselves make you wonder, if *you* are doing the talking, then whom are you talking *to*? Or, if you are listening, then who, precisely, is talking? Are we one person or two? Or many? Where does that voice we call "thought" come from? Who is the voice in your head, and how did it get there?

In the 1970s the Princeton psychologist and philosopher Julian Jaynes wrote a fascinating, bestselling book with the rather opaque title *The Origin of Consciousness in the Breakdown of the Bicameral Mind.* Jaynes's insights as a philosopher and a psychologist remain respected (he passed away in 1997), but the book was highly controversial. He speculated that consciousness of the kind I just described is an extremely recent evolutionary development. Between 10,000 B.C. and 1000 B.C., he argued, modern humans thought that the voice they heard in their head wasn't their own, but the voice of a chieftain or demon or god, some very real being outside their own minds. In other words, they didn't talk to themselves the way we do, they instead believed they were listening to another all-knowing being who was observing them and their thoughts. Jaynes called this kind of mind bicameral, or two-chambered, one chamber that listened and one that spoke, but neither of which was aware that they were part of the same brain. "[For bicameral humans], volition came as a voice that was in the nature of a neurological command," he wrote, "in which the command and the action were not separated, in which to hear was to obey." As evidence he points to the statues and idols that ancient cultures in Egypt, Sumeria, and Mesoamerica created as the physical symbols of the gods and chiefs that spoke to bicameral humans; the voices. This is the only possible explanation, he argued, for why they were built, and, having been built, why they exerted the enormous influence they clearly did over their cultures.

Jaynes also maintained that the people who lived in these ancient societies did not in any way believe others truly died. They only moved on to another world and then, having arrived, spoke from that world directly to those left behind. And from that world the gods also spoke, commanding the creation of great temples and the elaborate rituals for their benefit because, after all, they *were* running the show. The first laws, Jaynes explains, such as Hammurabi's Code and the Ten Commandments, were, as Hammurabi and Moses both said, rules passed directly to them by God.

Is Jaynes right? Were we once incapable of thinking for ourselves? Or more precisely, was there a time in our evolution when we were unaware that we *were* thinking for ourselves? It's impossible to know, absolutely, what was going on in the minds of humans living in ancient Sumeria, Egypt, or the Yucatán thousands of years ago, but we do know that the human brain, even today, sometimes struggles to identify the speaker within us as our "self." Schizophrenics hear a voice, or sometimes multiple voices, that they do not recognize as belonging to them. They come from "others" speaking from the outside. Yet brain scan studies illustrate that the voices are, in truth, being generated inside their own heads.[1]

The experience of schizophrenics, and Jaynes's theory, raises the question that somehow the human brain came up with a trick that helped it talk to, and control, itself. At some point it found a way to transform those external voices into internal ones. If that's true, though, then how did we manage it?

The answer begins with our brain's unique ability to create symbols and weave them together in outrageously complex ways. Other animals can't invoke symbols, but they can associate a single symbol or event with an experience they hold in their brain. Your dog, Fido, for example, may recognize the sound (but not the meaning) of the word *walk* and associate it with something he likes to do at the end of a leash with you each evening, but that's the extent of the connection between the two. The *sound* you make when you say "walk" brings to Fido's mind a specific experience, and so when he hears you make that noise, he runs for the door and waits. Scientists call this an iconic relationship between an experience and its external representation.

If you move a little farther along the evolutionary chain, you will find that primates possess more sophisticated symbolic capabilities. Take the case of two remarkable chimpanzees, named Sherman and Austin, at the Language Research Center at Georgia State University. Both were trained to associate specific symbolic pictures, or lexigrams, with certain events. The lexigrams were imprinted on a series of buttons in the laboratory where they trained. When they pressed a lexigram, they might find themselves rewarded with a goody, like a banana, or banana juice. Pretty quickly the lexigram came to stand for the reward in the minds of Austin and Sherman, not unlike the way the sound *walk* became linked to a good time outside for Fido.

Once the two chimps figured out the one-to-one relationship between a specific image and its reward, the researchers decided to present them with a new challenge. This time they were required to use two buttons in combination, like a verb and a noun, to receive a treat. One kind of lexigram represented "give" or "deliver," and the others represented a specific kind of food. So to receive a banana, Sherman and Austin had to hit the lexigram for "give" and then the correct one for "banana." This took some doing because there were multiple combinations of foods and commands, or verbs and nouns. But after some intense training, the chimps got the hang of the new system.

However, researchers had still more in store for the two hard-working chimps. Once they had absorbed the rules of the two-icon system, they were next provided a different alphabet of rewards and verb symbols that they had to use to receive treats. The question now became, could Sherman and Austin transfer the command-reward system they had learned to entirely new lexigrams on their own? After some trial and error, they again rose intrepidly to the challenge and learned the new system.

The big insight here is that they had comprehended an underlying organizing principle that made it easier to master the new lexigram vocabulary, an ability called an indexical symbolic relationship, one in which an animal can transfer a particular way of thinking to different situations. It represents a huge leap from iconic thinking because iconic symbolic relationships merely require one-on-one memorization.

If chimpanzees can manage this today, it's likely our direct ancestors going back millions of years could master something like it as well. So what makes us different from them? Our special ability is that we cannot only make iconic and indexical connections between meaning and experience like Fido and Austin and Sherman, we can weave indexes of symbols into much more intricate systems of entirely new symbols, and we can do it in an almost infinite variety of ways.

For example, you not only see the letter *e* in the words you are reading on this page and associate a sound with that letter (an iconic relationship), but your mind effortlessly combines many *e*'s and other letters into words, each of which has greater meaning than the individual sounds of the letters. And then you can pile together the words into sentences that have greater meaning than any one word, and so on. You also simultaneously understand the context of the letters. An *e* in one place can represent one sound, in another it can be silent (in English, anyhow).

Words can also change their meaning depending on the context, just as letters do. The interrogative "Turn right, right here, right?" uses the same word in one sentence three times, yet each time it has a different meaning because its context is different. Your mind understands this because it also grasps the underlying rules of the English language. It grasps them even if it can't entirely explain them.

Chimps may eventually and painstakingly be able to comprehend a simple system of language like "(You) throw ball"—subject, verb, and object. But there will never be the ghost of a chance that any chimp, even if he is the Shakespeare of *Pan troglodytes*, will comprehend the intricacies of the apparently simple question "Turn right, right here, right?"

One of the reasons he can't is that all human language is recursive, which is another way of saying that it can embed concepts within concepts. Just as letters are embedded inside words, strings of words can be embedded within sentences to make them more meaningful. Take the sentence "John, devilishly handsome as he was, refused the title King of the Prom, even though he secretly believed he deserved it." The ideas that John was devilishly handsome, that he secretly believed he deserved the title, and that the title was King of the Prom all invest the basic sentence "John refused the title" with much deeper meaning and a lot of useful information. They tell us about John, his motives, why he has the feelings he has, and provide some insight not only into how he looks, but how he looks at himself. Yet all of these additional nuggets of thought sit nicely nested, one inside the other. Unlike iconic and indexical meaning, this ability is richly symbolic, and unique to our kind of brain.

Language isn't the only symbolic system we have created that taps our special ability to recursively weave bundles of symbols into mosaics of meaning. We do it in mathematics with numbers and variables that can be built into proofs and theorems and formulas. In music we stack notes elaborately to construct melodies, then fashion melodies into themes and songs and symphonies, and even thread in harmonies with still other melodies, and finally, for good measure, add words to the melodies to create everything from pop hits to operas and Broadway shows.

Paintings are collaborations of symbols, too. Different colors of paint are situated together to represent real objects or feelings or ideas so that together they fashion works that have broader meaning than each of the dabs or drops of color themselves. Georges Seurat's pointillist paintings are a perfect example—hundreds of thousands of individual dots of different colors that together bring an image alive. Without our ability to gather together an alphabet of symbols and connect them in elaborate, nested patterns, there would be no *Hamlet* or *Faust* or *Moby-Dick*, no laws of thermodynamics, no science, music, architecture, no Kabuki theater, sculpture, Renaissance art, or anything else that has made the great, expansive construction project that we call human culture possible. Every iota of it is built on the unique and powerful talent of your *Homo sapiens* brain to mysteriously direct the molecular machinations of its own neurons to manufacture symbols and then share them with the other symbol-recognizing creatures around you. This enables minds to meld, hearts to bond, and ideas to be shared and bent and shaped by many other minds. And the brain does this without really comprehending how it does it, something like the way a basketball player drives to a basket and deftly deposits the ball into it without a moment's reflection.

We are able to embed symbols within symbols this way, and create intricate and outrageously complex thought structures, because our brain can take a concept, idea, or goal and set it aside temporarily while we shift our attention and work on something else. Scientists use two symbols to describe this earth-shattering capability: *working memory*.

In its simplest form, working memory is something like taking a call on your cell phone, starting a conversation, and then asking the caller to hang on while you take a second call. You can then begin that conversation without losing sight that you have another conversation-in-waiting because you have filed away the first phone call as a symbol, a kind of "object" the brain can retrieve from a folder or drawer. It's almost as though it is a physical thing. This holds true for nearly anything we can think or imagine, from goals to concepts to worries.

What's more, these "objects" we put on hold can have multiple concepts that live within *them*. We don't have to remember each of the individual pieces of information that reside within what we have set aside, we just have to be able to recall the big idea. And when we do, everything else comes along for the ride. If you are envisioning the Taj Mahal, you don't have to log and file away every detail while you shift your attention to preparing lunch. You simply prepare your meal, then reach back into your mind and pluck up the concept "Taj Mahal," and all of your thinking related to it returns like a nicely nested matryoshka doll, a file folder of the mind.

It's not clear where, precisely, this talent lies in our brain. Like nearly everything else cerebral, it almost certainly doesn't reside in one place. The brain, like recursion itself, is nested and networked, woven together. But fMRI studies have found that humans and other primates have regions called the frontal operculum, which activates when they process indexical kinds of information, but only humans have the much more recently evolved Broca's area, which handles language and grammar and syntax, recursive symbolic systems.

Broca's area is part of the human prefrontal cortex, a sector of the brain that sits directly behind our foreheads. It is one of the reasons we find ourselves with the large, childlike heads we have, and why our foreheads don't slope back as much as our cousin primates. The human prefrontal cortex (PFC) has, in evolutionary terms, sprinted toward its present state compared with other advances in brainware. While our brains have tripled in size over the past six to seven million years, our PFC has increased sixfold. When it comes to symbolic thinking, this is where the action is.

The action is here because the PFC has evolved as the brain's chief executive officer. It polices, as much as they can be policed, the primal, impulsive activities of our minds. The PFC inhibits anger, fear, hunger, sexual attraction, and other strong, but ancient drives. Many of these capabilities emerged as our ancestors became increasingly social and more reliant on one another for survival. Evolution would have favored individuals with brains that were better at controlling purely selfish impulses and favored those that took a longer-term view of situations. When you live in groups, after all, it may not pay to act solely in your short-term self-interest since you may need the help of others in the future. So today perhaps you share your food so that another day food may be shared with you.

Not only does the prefrontal cortex act as an executive this way, it allows us to think ahead by taking symbolized ideas, concepts, and memories and cobbling them together into scenarios that are completely nonexistent except in the brain itself, or, put another way, imagined. By recalling information held in long-term memory, packaging that with new information, setting these newly organized symbols aside in working memory so they can percolate while we move forward with other goals and ideas, we advance through the day, prioritizing, organizing, imagining, worrying, creating. Sometimes the work is mundane, like figuring out how to get showered, make phone calls, answer e-mail, and catch the subway in more or less the correct order so we don't show up at an appointment late, unbathed and misinformed. Sometimes the work is profound and results in the special theory of relativity. You never know.

It doesn't take much to imagine that the special abilities of the prefrontal cortex, whenever they finally came together, made our species dramatically different, outrageous really. Able to represent our thinking symbolically, then to embed these symbols inside one another, we became beings able to efficiently create, organize, and recall enormous amounts of complex information for still more revision.

In many ways it produced in us the same sort of ability that digital compression algorithms make possible. JPEG images and MP3 sound files, which make showing off the family photos on your iPad or listening to your favorite music on your phone possible, are the results of compression algorithms. What makes them useful is that they are not perfect reproductions of the images and sounds they represent, but the formulas that create the algorithms excel at extracting just enough of the *right* information to re-create a close facsimile that requires far less information and memory than the original. The copy is similar enough that most of us can't tell the difference, yet it's far less information intensive, and therefore more efficient. Symbols do the same thing. We don't remember everything. Just what we need to remember.

As helpful as all of this symbolization is in the business of expressing what you feel and think to others, it also makes it possible for you to do something else rather amazing: explain what you feel and think to yourself. In fact, it makes your "self" possible, and that may be the most stunning illusion our brains have pulled off yet over the past fifty thousand years—maybe the most stunning ever. The ability to create the ultimate symbol: you. Which takes us back to the question we asked at the beginning of this chapter: Who are "you," anyway?

When you are thinking, and talking, to yourself, the you that you are speaking to is a symbol. Like a reflection in a mirror, it is a representation made possible because your brain can generate symbols. Just as your mind symbolically represents the other people in your life, it also uses this trick to represent a version of you, which makes possible an enormously powerful force in your life, this second you, who is diligently and deeply influencing your every feeling, thought, and choice.

But to exert this influence, your brain pulls off still another astonishing feat. It changes itself physically. The symbolic "you" alters the real "you." Study after study has shown that the generation of our own thoughts and memories transforms the chemical and physical structures of our brains, in real time. Mind-boggling as we might find this, we shouldn't be horribly shocked. If our brains are the prime drivers of our behaviors, then when we think, feel, imagine, or change our minds in any way, our brain *must* also change. How can it not? Our actions, feelings, and thoughts are simply the dynamic reflections of the brain's physical, chemical, and electrical states. If you doubt that your brain dictates your reality, just drink a few shots of tequila. Your reality changes because your brain chemistry has been remixed. The same is true, if more subtly, when you whip up your brain chemistry by worrying or recalling a warm memory or losing your temper because that guy in the truck cut you off in traffic this morning.

In this way the brain changes itself, commands itself, reacts to itself, reshapes itself. It somehow bootstraps self-awareness and self-determination and simultaneously generates a symbolic self to be aware of and to command (as opposed to a god or a demon who dictates orders). This is a little like a box of cake ingredients purposefully opening and mixing themselves, then hopping into an oven to bake. This means that you and me (and every other of the seven billion humans currently alive on planet Earth) are, as Douglas Hofstadter might put it, "a strange loop," a supreme example of recursion, a matryoshka doll of selves.

To understand why we have come to operate this way, think of social interaction as a kind of rapidly changing ecosystem made up of a mix of personalities that requires constant adaptation to the shifting agendas, relationships, alliances, and power struggles within the group. In the highly social and very bright species that preceded us, part of the battle for individuals would have been to keep motives and relationships straight in their own minds. Those among our ancestors who could successfully track and recall the behaviors of their friends and enemies would have excelled, survived, and passed their genes along.

To manage this, they must have learned to symbolize different personalities. Maybe Goog tended to be aggressive; Targ, helpful and friendly; Moop, well organized and smart. This would have helped them "slot" others into organizational categories so they could deal with them in ways they saw fit, depending on their own personalities. Since these relationships only matter in so far as they are connected to you, along the way it would have been impossible not to eventually apply the same index to you. We became to ourselves another person in our social ecology.

As evolution continually favored smarter and increasingly self-aware creatures from *Homo erectus* to *ergaster* to *heidelbergensis*, Neanderthals, Denisovans, and *Homo sapiens* eventually emerged. Both we and Neanderthals developed large brains and complex prefrontal cortices, but we developed in different parts of the world, under entirely different circumstances, split from a common ancestor.* We both may have developed spoken language, but very different kinds. We were both self-aware and capable of symbolizing, but to what extent remains unclear. Neanderthals may never have developed a highly complex and fully symbolic inner world, and *Homo sapiens* may not have pulled off this level of cerebral legerdemain themselves until fifty thousand years ago, maybe later.

Perhaps then the prefrontal cortex reached a plateau where it could not only fully symbolize others, but manage the one last thing that made us so profoundly different from all other primates and humans that had come before us or even grown up with us: symbolize ourselves. With that, everything changed, radically. Because when we could fully symbolize ourselves, it meant that we could also begin to embed our symbolic selves among all the other symbols around us. We could begin, entirely inside our minds, to imagine what we would do before we did it. We could guide our behaviors, or at least conceive of guiding our behaviors, the way a chess player moves the pieces on a chessboard. By creating a symbol of ourselves, we became conscious and self-aware, capable of purposefully planning our behavior.

We could imagine.

That, in itself, represents a remarkable leap, but it made still another leap possible. The moment we consciously act on a scenario that we have imagined, it means we have taken control of our behavior. We have consciously made choices and acted on them. With the invention of a symbolic "you," intention and free will were born. Or at least the convincing illusions of them.

So the voice in your head that is talking to you? It's you. But the person that is listening isn't, not precisely. It's a symbol you have created. An image of you, a virtual version, like an avatar in a computer game, except the computer game is your inner mental life, the place where you map your actions and make your choices before transforming them into reality. The "you" you talk to is a simulation.

Looked at this way, our "selves" are, quite literally, a figment of our imaginations, the ultimate illusion, but an extremely useful one because this illusion has enabled us to take a hand in the control of our

---

*The word is still out on Denisovans and the Red Deer Cave people. Even *Homo floresiensis*. Denisovans appear to also have descended from *Homo heildelbergensis*.

fates, at least more than any other creature ever has. We are not only an animal that can explore a life not yet lived, and dream of a future we desire, we can also take hold of those dreams and make them come true. Out of a chaotic flux of random events in nature that have no agenda and are utterly incapable of making any plans, we have evolved into a planning, agenda-making, dream-conjuring creature. We are the first survival machines to also become living, breathing imagination machines.

If you compare us with other animals, our ability to create symbols turns out to be a kind of superpower, like being able to fly or peer through rock with X-ray eyes. They are super because they have transformed us into a world-changing, supremely adaptable, überbe-ing. Not simply "a figure in the landscape," as Jacob Bronowski put it, "but the shaper of the landscape . . . the ubiquitous animal." This isn't human arrogance speaking, declaring that the greatest advance in all of evolution has been the sudden emergence of our kind. This takes nothing away from the remarkable abilities of other animals. It simply states an irresistible fact, as sure as blue whales are gargantuan, chee-tahs are swift, and grunion dance on the beach under full moons; we alone have developed this superpower that lets us make symbols, and that has made us, bar none, the most adaptable creature planet Earth has yet witnessed.

The strange thing is, this virtual version, this symbol of our "selves," not only reflects on itself (and selves within selves like the reflec-tions of two facing mirrors), but also reflects on the mind that makes it possible in the first place. That can sometimes make our mental lives even more complicated than they already are. Superpowers, it seems, often bring difficult trade-offs in tow. Like mental illness, for example.

When I was a child, I once asked my mother why our new Chrysler New Yorker didn't have power windows, when they were all the rage in the newest cars. "The more fancy the gadgets," she answered, "the more there is to break down. We would only have to fix it later." The makers of intricate technologies like cars, computers, and spaceships have inevitably found that when the engineering of anything reaches a certain level of complexity, it is much more difficult to maintain than simpler systems. A straw rarely breaks down. Nor do paperweights, generally. Space shuttles, on the other hand, were designed with thousands of "redundant systems" because so many functions could go haywire. Yet when it comes to complexity, a shuttle holds not even the dimmest candle to the human brain. The brains you and I carry into adulthood have an estimated 100,000,000,000 (one hundred billion) neurons, each connected to as many as a thousand other neurons. This is sophistication of the incomprehensible variety. The human brain is so convoluted in its wiring, genetics, and neurochemistry that it is a wonder that so many of them function so well. Of course those that evolved and didn't work up to snuff quickly resulted in the death of their owner and were swiftly tossed from the gene pool. Still the modern human brain will, and does, often go sideways. We see it in the number of people who suffer from chronic depression, not to mention more dramatic and damaging conditions such as bipolar disorder, schizophrenia, autism, obsession, compulsion, attention deficit, and multiple personality disorder. And that's just some of the labels we use to describe mental illness. The more we come to understand the human brain, the more we discover what can go wrong with it. Arguably, no human brain is really "normal."

Mental illnesses like these are uniquely human because they are linked to uniquely human capabilities like language, symbolization, and working memory. Neurologists and psychologists have been especially curious about two mental illnesses—schizophrenia and autism—and what each has to say about the way we evolved.

Schizophrenia is a mental disorder characterized by trouble differentiating between reality and imaginary worlds and experiences. Schizophrenics can suffer from extreme paranoia, delusions, disorganized speech and thinking. In severe cases they often hear voices and many times carry on conversations with the voices because to them they aren't the voice most of us identify as our "self," but belong to someone else, usually unseen. This is something like Julian Jaynes's bicameral mind. Those who speak are distinct, often conflicting, even abusive, and it can be maddening, or, occasionally, captivating. One schizophrenic reported waking up to hear two Israeli generals debating battle strategy, a subject he had never before in his life contemplated. "It was a fascinating experience," he recalled. It's an insight into the power of the human brain that someone can have that much detailed knowledge about a subject as arcane and complex as military strategy, yet not consciously realize it. It makes you wonder what treasures of information lie untapped within each of our minds.[2]

Schizophrenics suffering from acute paranoia often convince themselves that they are being hunted or persecuted. These fears can become part of an elaborate imaginary world they live in, a world that is as absolutely real to them as our daily lives are to the rest of us. The richness of these worlds and scenarios are further testaments to the power of human creativity. We on the outside see this as madness, and it is excruciating for those who suffer through such horrible conflicts and feelings, but it is possible to see how our minds can go to places like these. After all, don't each of us hear a voice that often sends us conflicting messages? The only difference is that we identify that voice as our own, not as belonging to unexpected, intruding, and ethereal companions who pop into our consciousness unbidden and unannounced.

And don't we all live in imaginary worlds of our own making—in the tomorrows that we plan; the lives that we lay out; the conversations we imagine having with friends or enemies? Every piece of fiction ever written is an elaborate imagining manufactured out of the symbols in the mind of its author, no less labyrinthine in its way than the delusions of schizophrenia. The line between normalcy and madness may be finer than any of us would like to believe.

Autism is not usually as dramatic or debilitating as schizophrenia can be, but it, too, provides a glimpse into the mysteries of the creative spirit. Like schizophrenia, autism runs along a spectrum from mild to severe, and some of the underlying symptoms for it are similar: difficulty socializing with others, a tendency to become obsessed with specific behaviors, sometimes self-injury or the need for repetitive rituals that might involve entertainment, food, or dress. In about one case out of ten, autistic people develop remarkable talents, but are otherwise incapable of leading what the rest of us might consider a normal life. Researchers sometimes refer to them as autistic savants. The movie *Rain Man* was based on real-life autistic Bill Sackter and another savant, Kim Peek, who, for reasons not entirely clear, was blessed with an astounding memory, yet struggled with some of life's most basic undertakings. Sackter passed away in 1983, and Peek died of a heart attack in 2009, but both were remarkable people. Sackter was also a model for Charlie Gordon in the novel *Flowers for Algernon* and the movie it inspired about a mentally retarded man who becomes a genius before returning to his earlier state. Peek could read thousands of pages of facts and trivia, then much later recall with almost perfect accuracy the information on those pages, say, the weather on December 14, 1964, or Roberto Clemente's batting average in 1967.

Other autistic savants have been genetically bestowed with extraordinary talents as musicians, painters, mathematicians, sculptors, even writers. Sometimes the talents are wide-ranging and accompanied by high intelligence, as in the case of Matt Savage, who at age six taught himself to read piano music and went on to study both jazz and classical piano at the New England Conservatory of Music. In between he also somehow found time to win a statewide geography bee, compose many of his own pieces, and release nine albums while touring the world and appearing with an impressive list of jazz greats.

Alonzo Clemons on the other hand has an IQ of 50, the result of a severe brain injury as a child. Strangely, though, Alonzo developed a talent for creating marvelously accurate animal sculptures out of clay, even if he had only caught a glimpse of the animal or seen a photo or drawing of it in two dimensions. His works have sold for tens of thousands of dollars. When looking at Clemons's works, it's difficult not to think of the fluid, breathtaking artwork in the caves of Altamira and Lascaux. Were these the works of a Cro-Magnon savant, someone seemingly endowed with magical talents, and magical ways of representing the world?

Seth F. Henriett is another savant blessed with a high IQ like Matthew Savage, and a marvelously broad array of talents. Though Henriett suffered from severe social problems and autoimmune disorders early in her life, she also revealed aptitudes for music and art. She played flute at the age of seven and contrabass at the age of eleven. By age thirteen her abstract and surrealistic paintings were gaining attention. Shortly afterward she wrote two books about her experience as an autistic and won several international writing competitions with her stories, essays, and poems.

Despite these staggering skills, each of these people has difficulty relating to others. They shun making eye contact or being touched; they often prefer to be alone and struggle with even basic personal interaction. Yet don't all of us exhibit a quirk or two, or three? Phobias, preferences, habits, interests, even obsessions? Various experts have speculated that some well-known people in history were autistic to some extent, or another, including Lewis Carroll, Charles Darwin, Emily Dickinson, Thomas Jefferson, Isaac Newton, and Wolfgang

Mozart. How much emptier would human civilization be had it been bereft of the genius of these minds?

On the other hand, 90 percent of autistics do not become savants, though a significant number are highly functional. Again the affliction is not binary, one or the other, all on or all off. You and I might have smidgens of autism and not realize it, especially if you happen to be a man. Scientists have sometimes described autism as an extreme version of the male brain. And in truth, of all the world's autistics, only one fifth are female.[3] This may be because women have more axons and dendrites, which are the pathways in the brain that enable it to work as a unit. Men's brains have more neurons. In effect, this makes male brains less networked than women's, but outfitted with more processing power, largely focused, it seems, on spatial and temporal capabilities. This doesn't make one sex smarter or more talented than the other, simply different. It also helps explain, at least according to some scientists, why men are sometimes less socially tuned in than females, and why women are superior, generally, at reading social cues.

The issue with autistics isn't so much with the number or the function of their neurons, but that they suffer from a dearth of connections between them. What would cause this scarcity? Neoteny, or more accurately the processes that make neoteny possible.

Remember, complexity creates more opportunity for something to go wrong. During the crucial first three years that follow birth, when the brain triples in size and personal experience so strongly shapes the billions of pathways between neurons in the brain, cerebral development may mysteriously go awry in people who grow up autistic. Connections may be delayed, accelerated, or stunted. Studies show that different sectors of the brain develop in ways that keep them more separated from one another than normal, like islands in a sea, out of touch and segregated. Yet, some modules may become overwired, which could help explain remarkable feats of memory, mathematics, music, or art and views of the world so different from those of the rest of us.

The downside, of course, is that the segregations also make it more difficult to be socially sensitive and tuned in to other people's nonverbal communications—their smiles, tones of voice, body language— the little things we unconsciously and effortlessly do that grease the skids of human relationships. These create a deficit of a Theory of

Mind, a brain largely incapable of symbolizing its owner as a self to itself, let alone symbolize others. Rather than symbolization going rampant and boisterous as it does in schizophrenia, in autism it is reduced, stunted, and balkanized, with the manifold human genius for socialization sometimes being abandoned in exchange for a single, condensed, but breathtaking talent.

Why, if evolution so ruthlessly discards traits and behaviors that undercut a living thing's ability to survive and mate, have mental illnesses like these and others survived? Can they serve a purpose? Or did they once? In a study published in *Nature* in 2007 researchers led by Bernard Crespi and Steve Dorus analyzed human DNA from populations around the world as well as primate genomes dating back to the shared ancestor of both humans and chimpanzees to get a handle on what genes led to schizophrenia, why they evolved, and why the illness is still among us. They were astonished to find that of seventy-six gene variations known to be strongly related to schizophrenia, twenty-eight showed sturdy evidence that they were favored by natural selection when compared with other genes, even those associated with the most severe forms of schizophrenia. In other words, the genes weren't randomly repeated accidents; the forces of evolution were actively selecting them and passing them on. Why?

It could be that they are bound to other genetic talents that are extremely important to human survival, like speech and creativity, for example. One current theory is that schizophrenia is a "disorder of language" that represents a trade-off some *Homo sapiens* made in exchange for the remarkable gift of speech and consciousness the rest of us enjoy. Says Crespi, "You can think of schizophrenics as paying the price of all the cognitive and language skills that humans have." That may explain why 1 percent of the human race suffers from some form of schizophrenia.

Multiple theories connect schizophrenia and autism to the evolution of our ability to symbolize others and model their behavior in our minds; and our unique talent for symbolizing ourselves by using systems like language to talk to ourselves, imagine what others are thinking and intend, and envision events that haven't yet happened and never may.

Individually, the origins of both illnesses may be the result of brain development in childhood that misfires. Since the prefrontal cortex is, ultimately, a consequence of neoteny, the precise timing that the development of a modern human brain requires may be the source of both disorders. Some scientists have speculated that in schizophrenics, neoteny is retarded, or processes it sets in motion aren't completed. It's interesting that in most schizophrenics severe symptoms don't show themselves until around eighteen years of age, or older, when the brain has placed the majority of its design in order.

In the case of autism the complex cerebral structures and relationships that make Theory of Mind, language, and symbolization possible could all be affected early in development. We already know the brains of autistic children grow significantly faster and larger than normal between the ages of one and sixteen months of age and remain larger until ages three to four. Researchers have also found that children with autism develop 67 percent more neurons in their prefrontal cortex and have heavier brains for their age compared to typically developing children. It's almost as if connections made prior to birth and early in life arrive before they can be properly deployed.

With both illnesses something like a genetic wrench seems to have been thrown into the complex developmental processes that construct the foundations of a human mind during those long childhoods that neoteny has made possible. The timing and expression of genes that catalyze the cerebral alchemy of human behavior falter somehow, and once they do, it changes the brain in ways that aren't easily repaired, at least not based on what we know today.

Still there is a larger point to all of this. Mental illness, a state in which the mind is unable to get a solid handle on what the rest of us generally agree is "real," could not exist until nature first created a brain that could model the thing we call reality in the first place. That means it takes a human mind to suffer mentally. Cats, dogs, and other primates may endure depression or grow sad, they may develop lifelong fears and strong addictions, but they don't hear voices, imagine alternate realities, or suffer from an inability to speak or empathize. And they don't because they never enjoyed those capabilities in the first place and never will.

Further advances in genetics and brain imaging may reveal exactly how mental illnesses like these work and in the process expose to us some of the slick tricks the brain plays to create the illusions of self and reality. These advances already hint that the borders between reality and delusion are slim. Or more accurately, that reality *is* an delusion, just an extremely useful one. In some ways the brain is like the Wizard of Oz, standing hidden behind a curtain, spinning the wheels and operating the levers that create the illusory symbols that make our "I' and our reality possible.

All of this happens because of the elegant physical, pharmacological, and electrical interactions of the three pounds or so of wetware you currently tote around in your skull. These trillions of cerebral interactions know nothing of jealousy, love, passion, creativity, or sadness, and yet out of them emerge the threaded experiences we perceive as ourselves living a life connected, to varying degrees, to all the other humans that we encounter every day, day in and day out, until the brain that makes it all possible finally ceases to function.

Once the human brain materialized in the form that we now know, outfitted with its genius for creating and shuffling the symbols that make language, imaginary worlds, and, above all, that phenomenon Hofstadter calls the "anatomically invisible, terribly murky thing called I," creatures emerged that could dream, act on their dreams, and share them with the other "I's" around them. And that changed the world.

Our special talent isn't simply that we can conjure symbols or even weave elaborate, illusory tapestries of them, but that we can share these with one another, roping together both our "selves" and our imaginings, linking uncounted minds into rambunctious networks where thoughts and insights, feelings and emotions, breed still more ideas to be further shared. Creativity is contagious this way, and once a light emerged, it must have gone off like fireworks.

This has made every human a kind of neuron in a vast brain of humans, jabbering and bristling with creativity, pooling, pulling, and bonding ideas into that elaborate, rambling edifice we have come to call human civilization. In this way *memes*★ have traveled along the transit lines of our relationships, some of them seeing their way to reality, others run aground for lack of interest or use, selected out and gone extinct as surely as dodos, dinosaurs, and the flightless crake.

The wheel is a great example of a meme. So are the arch and the soufflé and a catchy tune like "I've Got a Lovely Bunch of Coconuts," or plumbing, sanitation, myths, and the Pythagorean theorem. Once

---

★A term coined by evolutionary biologist Richard Dawkins in his book *The Selfish Gene.*

upon a time, someone conjured up a large, circular thing that helped move heavy loads when paired with other similar circular things, and the idea stuck and was shared—wheel!

As memes spread, they mutate and combine with other memes and snap together in our social worlds as neatly as genes do in the molecular one. They breed only because we humans have the ability to both conceive and duplicate them, using all of the symbols we so ardently exchange.

The voice we had begun to hear in our heads—perhaps fifty thousand years ago—was a harbinger and a catalyst, the first step we needed to take before we could attempt to direct our own lives, fashion memes, and then transform them (for better or worse) into reality. At first the sharing must have been slow. It takes time for ideas to be passed along and built upon in a world where only a few tens of thousands of symbol-making creatures live. Still, compared with the geologic and genetic measures of time that preceded it, these changes came swiftly, and they gathered speed exponentially. Inside of forty thousand years agriculture and animal domestication were widely adopted, then came settlements, villages, and towns. Cities in Mesopotamia and the Middle East arose a mere nine thousand years ago. Despite wars, famine, disease, and natural disasters, we have surged forward since, inventing science, a global economy, vast communication systems that exchange thick streams of media, immensely complex governments and businesses, all of them, in their way, ceaselessly shuttling the proliferating agglomerations of human thinking around the globe every day—a humming and titanic network consisting of seven billion symbol-makers busily exchanging their symbols. Thanks to us they move from one mind to another as surely as genes move and mutate from one person to another.

Was the emergence of a brain—shaped in childhood and capable of symbolizing its owner—the final piece in the human puzzle, the last brick that completed the construction of what we today call truly "human"? Was this the evolutionary act that made civilization possible? We will never know for certain because we weren't around at the moment of modern human awakening.

Trying to figure out how that white light of the first symbolic insight came together is a lot like reverse engineering some alien engine we have found in a desert, fully operational, but without a manual. My guess is we will never fully comprehend how we turned the corner to become the human beings we are today. The brain itself may be the issue. Maybe the mind that it makes possible will always find itself just short of grasping how it creates its illusions, or why. Too much is at work in the unconscious, too much unavailable mystery. It doesn't mean we can't try, though, the way phycisists have tried to approach absolute zero. By definition it is impossible to get to a place where there is nothing, but we can keep working to get closer. As with the quest to reach absolute zero, maybe we can only ride upon the illusions it conjures and see where they lead. At least until a new kind of human evolves.

EPILOGUE: THE NEXT HUMAN

The need is not really for more brains, the need is now for a gentler,
a more tolerant people than those who won for us against the ice,
the tiger and the bear. The hand that hefted the ax, out of some
old blind allegiance to the past, fondles the machine gun as lovingly. It is a
habit man will have to break to survive, but the roots go very deep.
—Loren Eiseley, *The Immense Journey*

In THE BAY of Naples, not far from the shadow of Vesuvius, swim two seemingly unremarkable creatures, a common sea slug, and a jellyfish, called a medusa. The jellyfish, researchers know, carelessly bob through the upper waters of the bay and after birth quickly mature into full-grown, elegant adults. The sea slug's larvae also contentedly ride the water's currents, apparently happy to live the lives that such snails generally do. You wouldn't think that these creatures could possibly have anything to do with one another, but it turns out they are intimately and strangely connected.

Marine biologists first saw this connection when they noticed that full-grown versions of the snails had a small, vestigial parasite attached near their mouth. Nothing you would immediately notice. But when they did notice it and looked further into the whole affair, they made their remarkable discovery. It seems that as the slug larvae bob through the bay, they often become entangled in the tentacles of the medusa and are then swallowed up into its umbrella-shaped body. At this point you would assume that the larvae would soon be done in as nice morsels by the predator jellyfish, but, that is not how it goes. Instead, and astonishingly, it is the snails that begin to dine, voraciously—first on the radial canals of the jellyfish, then on the borders of the rim, and finally on the tentacles themselves, until the medusa disappears altogether and is replaced by rather a large slug, with the small bud of a parasite attached to its skin right near its mouth just like the one scientists found in their research. The parasite had become the host, and vice-versa.

Lewis Thomas, the fine physician, researcher, and essayist, told this story in his wonderful 1970s book *The Medusa and the Snail* to illustrate how peculiar and connected life on earth is. It certainly does that, but I'm recounting it here because in the eccentric relationship of these two creatures lies the echo of what the future holds in store for the human race.

Usually at this point in a book like this, the inevitable—and heavily loaded—question arises, what next? Where will human evolution take us now after our long and astonishing adventures? Are we still evolving? And if we are, what will the next act look like? Can we expect ever-enlarging brains to cram themselves behind alienlike foreheads? Or will our noggins contract to the size of a walnut, shrunk down by media overload and pharmaceuticals (the dimensions of the human brain have diminished 10 percent over the past thirty thousand years)? Or perhaps we will grow weak, fat, and small of limb, vaguely resembling Jabba the Hutt, while simultaneously sprouting an extra digit or two to better handle all of the texting we do? It might even be possible, as one scientist has speculated, that we will diverge into two subspecies, one fit and beautiful and the other overweight and slovenly, a kind of real-world version of the Eloi and Morlocks in H. G. Wells's *Time Machine*, except without the cannibalism and enslavement, we hope.★

Evolution, as the past four billion years have repeatedly illustrated, holds an endless supply of tricks up its long and ancient sleeve. Anything is possible, given enough millennia. Inevitably the forces of natural selection will require us to branch out into differentiated versions of our current selves, like so many Galápagos finches . . . assuming, that is, that we have enough time to leave our evolution to our genes. We won't, though, and none of these scenarios will come to

---

★This is the hypothesis of evolutionary theorist Oliver Curry of the London School of Economics.

pass. Instead, we will come to an end, and rather soon. We may be the last apes standing, but we won't be standing for long.

A startling thought, this, but all of the gears and levers of evolution indicate that when we became the symbolic creature, an animal capable of ardently transforming fired synapses into decisions, choices, art, and invention, we simultaneously caught ourselves in our own crosshairs. Because with these deft and purposeful powers, we also devised a new kind of evolution, the cultural variety, driven by creativity and invention. So began a long string of social, cultural, and technological leaps unencumbered by old biological apparatuses such as proteins and molecules.

At first glance you might think that this would be a boon to our kind. How better to better our lot than with fire and wheels, steam engines, automobiles, fast food, satellites, computers, cell phones, and robots, not to mention mathematics, money, art, and literature, each conspicuously designed to reduce work and improve the quality of our lives. But it turns out not to be that simple. Improvements sometimes have unintended consequences. With the execution of every bright new idea it seems we find ourselves instantly in need of still newer solutions that only seem to make the world more kerfuffled. We are ginning up so much change, fashioning thingamabobs, weaponry, pollutants, and complexity in general, so swiftly, that as creatures genetically bred to a planet quite recently bereft of technical and cultural convolutions, we are having an exceedingly difficult time keeping up, even though we are the agents of the very change that is throttling us. The consequence of our incessant innovating is that it has led us inevitably, paradoxically, irrevocably, to invent a world for which we are altogether ill fit. We have become medusae to our own snails, devouring ourselves nearly out of existence. The irony of this is Shakespearean in its depth and breadth. In ourselves we may finally have met our match: an evolutionary force to which even we cannot adapt.

We are undoing ourselves because the old baggage of our evolution impels us to. We already know that every animal wants power over its environment and does its level best to gain it. Our DNA demands survival. It is just that the neoteny that has made us the Swiss Army knife of creatures, and the last ape standing has only amplified, not replaced, the primal drives of the animals we once were. Fear, rage, and appetites that cry for instant gratification are still very much with us. That combination of our powers of invention and our ancient needs will, I suspect, soon carry us off from the grand emporium of living things.

The best evidence that we are growing ragged at the hands of the Brave New World we have busily been rolling off the assembly line is the growing numbers of us who freely admit to being thoroughly stressed. A recent study reported that the United States is "a nation at a critical crossroads when it comes to stress and health."★ Americans are caught in a vicious cycle: managing stress in unhealthy ways while assembling insurmountable barriers that prevent them from revising their behavior to undo the damage they are inflecting on themselves. As a result, 68 percent of the population is overweight. Almost 34 percent are obese. (This is rarely a problem in hunter-gatherer cultures.) Three in ten Americans say they are depressed, with depression most prevalent between the ages of forty-five and sixty-five. Forty-two percent report being irritable or angry, and 39 percent nervous or anxious. Gen Xers and so-called Millenniums admit to being more stressed about personal relationships than even their baby-boomer parents. It's so bad that the results of our anxieties have found their way into dental offices, where dentists now spend far more of their time treating patients for jaw pain, receding gums, and worn teeth than they did thirty years ago. Why? Because we are tense and anxious, grinding our teeth down to nubs as we sleep.

Stress, as the experience of lab rats everywhere has repeatedly testified, is a sign that a living thing is growing increasingly unfit for the world in which it lives, and as Darwin and Alfred Russel Wallace astutely observed more than 150 years ago, when a living thing and its environment are no longer a good match, something has to give, and it is *always* the living thing.

How are we handling our stress? Not too well. Rather than relaxing or getting more exercise when pressures mount, studies show that we instead skip meals, spend more time online or in front of the TV, then overeat and lie awake at night perfectly prepared to enter the next day bleary-eyed, short-tempered, and exhausted. What triggers this behavior? Those old primal drives and appetites we struggle so mightily to ignore.

Which returns us to the question, what next?

Our demise doesn't have to be a Terminator-style annihilation that leaves the world emptied of all humans, postapocalyptic cities stark and decaying with the smashed remains of our cultural accomplishments. It

★Conducted by the American Psychological Association, 2010.

may be more of a butterfly-like metamorphosis, a transformation in which we step over the Rubicon of our old selves and emerge as a new creature built on our own backs without ever realizing, at least early on, that we are no longer the species we thought we were. Did the first Neanderthal know that he, or she, was no longer *Homo heidelbergensis?* Those passages are made gradually.

Perhaps we will simply morph into *Cyber sapiens,*★ a new human, infinitely more intelligent than you or I are, perhaps more socially adept, or at least able to juggle large tribes of friends, acquaintances, and business associates with the skill of a circus performer. A creature more capable of keeping up with the change it generates. To handle the challenges of time shortages and long distances, *Cyber sapiens* may even be able to bilocate or split off multiple, digital versions of themselves, each of whom can blithely live separate lives and then periodically rejoin their various digital selves so that they become a supersize version of a single person. Imagine being able, unlike Robert Frost's traveler in his poem "The Road Not Taken," to choose both paths, each with a separate version of yourself. It makes you wonder if something essential in us might disappear should such possibilities come to pass. But then, perhaps, that is what will make the new species new.

A whole group of *Homo sapiens* are already contemplating what the next version of us might be like. They call themselves transhumanists, anticipating a time when future anthropologists will have looked back on us as a species that had a nice run, but didn't make it all the way to the future present. Transhumanists foresee a time when beings will emerge who will literally be part biology and part machine. In this I suspect they are right, the logical next step in a long trend. We are already part and parcel of our technologies after all. When was the last time you checked your cell phone or simply walked to work, hunter-gatherer style? We have long been coevolving with our tools. It's just that now the lines between humans and machines, reality and virtuality, biology and technology, seem to have become especially blurry and will soon twitch and blink away completely.

Transhumanists predict that by melding molecule-size nanomachines with old-fashioned, carbon-made DNA the next humans might not only speed up their minds and multiply their "selves," but boost

---

★A term coined in my previous book *Thumbs, Toes, and Tears: And Other Traits That Make Us Human.*

their speed, strength, and creativity, conceiving and inventing hyper-intelligently while they range the world, the solar system, and, in time, the galaxy. In the not-distant future we may trade in the blood that biological evolution has so cunningly crafted over hundreds of millions of years for artificial hemoglobin. We may exchange our current brand of neurons for nanomanufactured digital varieties, find ways to remake our bodies so that we are forever fresh and beautiful, and do away with disease so that death itself finally takes a holiday. The terms *male* and *female* may even become passé. To put it simply, a lack of biological constraint may become *the* defining trait of the next human.

There could be a downside to these sorts of alterations, I suppose, should we find ourselves with what amounts to superhuman powers, but still burdened by our primal luggage. Our newfound capabilities might become more than we can handle. Will we evolve into some version of comic-book heroes and villains, clashing mythically and with terrible consequences? Powers like these give the term *cutting edge* a new and lethal meaning. And what of those who don't have access to all of the fresh, amplifying technologies? Should we guard against a world of super-haves and super-have-nots? It is these sides of the equation I wonder about most.

Given evolution's trajectory, short of another asteroid collision or global cataclysm, we will almost certainly become augmented versions of our current models. That has been the trend for seven million years. Apes increasingly endowed with more intelligence, and more tools, becoming simultaneously wiser and more lethal. The question now is, can we survive ourselves? Can we even manage to become the next human? It's a close question.

I'm counting on the child in us to bail us out, the part that loves to meander and play, go down blind alleys, fancy the impossible, and wonder why. It is the impractical, flexible part we can't afford to lose in the transition because it makes us free in ways that no other animal can be—fallible and supple and inventive. It's the part that has gotten us this far. Maybe it will work for the next human, too.

ACKNOWLEDGMENTS

Sitting here at my desk on a warm morning in 2012, it's easy to think of book writing as an entirely private undertaking. Lots of time spent tapping away all by yourself, wrestling with sentences that refuse to make sense, wrangling obstinate phrases; lots of reading, library excursions, and Web-crawling, too, for obscure facts; and a fair amount of frantic note-jotting punctuated with mindless gazings out the window. Occasionally some solitary head-banging was known to go on.

But mostly the sequestered nature of writing is an illusion. A book like *Last Ape Standing* could never find its way into the world without the help and support of battalions of people. For starters, I've been privileged over the years to sit in long conversations with scores of scientists and some of the finest minds I've ever come across—Michael Gazzaniga, Gerald Edelman, Hans Moravec, Ray Kurzweil, Michael McElroy, and the late and remarkable Lynn Margulis, to name a handful. Those conversations provided me with perspectives for this book that I would never have developed without the benefit of their hospitality, capacious intellects, and experience. "But look at it this way . . ." is a phrase I heard often from all of them as they gently helped me out of the small seat of my limited perspective.

Then there are the hundreds of books, articles, and scientific papers I communed with for this project. Each represented years of work and research on the part of their authors, nicely condensed explorations of the corners of human evolution and behavior that I couldn't hope to cover in a thousand lifetimes. Whether it was global climate, human genetics, evolutionary psychology, anatomy, or history, the work of these researchers and writers provided the essential vitamins and minerals of the pages that follow. I may not know all of them personally, but I am deeply indebted to each of them.

Jen Szymanski and Frank Harris also deserve my special thanks. Frank for his excellent eye and artwork; Jen for her good nature, unfailing attention to detail, and unshakable reliability.

Books also don't happen without a publisher and an editor who believe in the book and are willing to accept that an idea can be transformed into something people will want to buy and read. I will be forever and deeply grateful for the insight and generous spirit of George Gibson at Walker/Bloomsbury. In baseball there are "player" managers. George is a "writer's" publisher, always encouraging, never negative, a fine example of the best that human evolution makes possible. This book's editor, Jacqueline Johnson, is perhaps the calmest person I have ever met, and no matter what sentences might be flailing around under my hand, what participles might be dangling, what questions I might throw her way, or what deadlines I bent and mangled, she remained as imperturbable as Kilimanjaro. And once again I owe my agent, Peter Sawyer, my deep gratitude for his excellent advice and insights on nearly everything, and for hanging in my corner no matter how cockamamy the ideas are that I bring to his sage ear.

Mostly, though, I owe my gratitude to my family. My daughters, Molly and Hannah, who have, their entire lives, put up with this thing I do, which Molly, when she was two years old, once described to someone as "hitting buttons." Their smiles and laughter and company place even the toughest days at the desk in perspective. My stepchildren, Steven and Ann, have gamely learned how strange it is to have a writer in the house, and still they haven't punished me for it or given me up for mad. Above all, though, I thank Cyndy, my incomparable wife and the best human on earth, for her relentless patience, encouragement, and love. How mercilessly I have sometimes bent her beautiful ear, yet we remain married.

NOTES

INTRODUCTION

1. Until recently, paleoanthropologists referred to the subfamily of homi-
   noids that consisted of humans and their ancestors as hominids, but even
   the convoluted argot of science sometimes changes. Today *hominid* refers to
   all great apes, including gorillas and chimpanzees, but *hominin* refers specifically to ancient and modern humans who split off from a common
   chimp ancestor seven million years ago, or thereabouts. These include all
   of the Homo species (*Homo sapiens*, *H. ergaster*, *H. rudolfensis*, for example),
   the australopithecines (*Australopithicus africanus*, *A. boisei*, etc.), and other
   ancient forms such as *Paranthropus* and *Ardipithecus*. The important point is
   that we are the last surviving hominin on earth.
2. During the writing of this book, two new modern human species were
   discovered and two move ancient species. For more on these, read the side-
   bars on pages 90–93, "The Newest member of the Human Family."

I: THE BATTLE FOR SURVIVAL

1. Some scientists have speculated that *tchadensis* and others like him at this
   time in prehistory could be the offspring of early humans and chimps who
   mated and brought hybrid "humanzees" into the world in the same way
   mating female horses and male donkeys conceive mules. Since such a hybrid wouldn't have been able to produce children of its own, the chances of
   this rare fossil surviving until the present are exceedingly slim, but in the
   world of paleoanthropology nearly anything is proving to be possible. For
   more, read "Human, Chimp Ancestors May Have Mated, DNA Suggests,"
   National Geographic News, May 17, 2006, http://news.nationalgeographic
   .com/news/2006/05/humans-chimps.html.

2. See chapter 1 of *Thumbs, Toes, and Tears: And Other Traits That Make Us Human.*

3. For more on this read, "Unlocking the Secrets of Longevity Genes," *Scientific American*, December 2006.

4. Additional information on this interesting theory can be found in "How Dietary Restriction Catalyzed the Evolution of the Human Brain," *Medical Hypotheses*, February 19, 2007.

2: THE INVENTION OF CHILDHOOD
(OR WHY IT HURTS TO HAVE A BABY)

1. This is an apt analogue for the situation our species faces today . . . our own intelligence has put us in a precarious situation that we, too, may not survive. See the epilogue, "The Next Human."

2. In one of his more whimsical essays written more than thirty years ago, evolutionary theorist Stephen Jay Gould depicted Mickey Mouse as a perfect example of neoteny in action. The older Mickey got, Gould pointed out, the younger (and cuter) his animators made him look. As he aged, Mickey acquired greater youth. Broadly speaking this is precisely what happened to the line of humans who eventually led to you and me.

3. See L. Bolk, "On the Problem of Anthropogenesis," *Proc.Section Sciences Kon. Akad.Wetens.* (Amsterdam) 29 (1926): 465–75.

4. "In neoteny rates of development slow down and juvenile stages of ancestors become adult features of descendants. Many central features of our anatomy link us with the fetal and juvenile stages of [nonhuman] primates." Gould, *Ontogeny and Phylogeny*, 1977, 333.

5. From Barry Bogin, "Evolutionary Hypotheses for Human Childhood," *Yearbook of Physical Anthropology* (1997), 70. "In Shea's view, a variety of heterochronic processes are responsible for human evolution. The others may be hypermorphosis, acceleration (defined as an increase in the rate of growth or development), and hypomorphosis (defined as a delay in growth with no delay in the age at maturation) . . . None of these acting as a single process can produce the human adult size and shape from the human infant size and shape. The same holds true for acceleration and hypomorphosis. In agreement with Schultz, Shea states that 'we [humans] have extended all of our life history periods, not merely the embryonic or juvenile ones' (pp. 84–5). Humans have also altered rates of growth from those found in other primates and possible ancestors. To accomplish all this required, in Shea's view, several genetic changes or adjustments during human evolution. Since the hormones that regulate growth and development are, virtually, direct products of DNA activity, Shea proposes that the best place to look for evidence of the evolution of ontogeny is in the action of the endocrine system. According to Shea and others (e.g., Bogin, 1988) differences in endocrine action between humans and other primates negate neoteny or

hypermorphosis as unitary processes and instead argue for a multiprocess model for human evolution."

6. Around this time, our ancestors may also have begun to lose their hair, another neotenic trait, although loss of hair almost certainly also helped to avoid overheating on Africa's scorching savannas.

7. Martin, "Human Brain Evolution in an Ecological Context" (fifty-second James Arthur Lecture, American Museum of Natural History, New York, 1983).

8. The Pleistocene epoch lasted from about 2.5 million years to 11,700 years ago and includes Earth's recent period of repeated ice ages. The Pleistocene is the first epoch of the Quaternary Period or sixth epoch of the Cenozoic Era. The end of the Pleistocene corresponds with the end of the last glacial period, the one that immediately preceded the blossoming of recorded human history. It also corresponds with the end of the Paleolithic age used in archaeology.

9. Another reason that children need a special high-energy diet is the rapid growth of their brain. In their research in 1992, Leonard and Robertson estimated that due to this accelerated growth, "a human child under the age of 5 years uses 40–85 percent of resting metabolism to maintain his/her brain [adults use 16–25 percent]. Therefore, the consequences of even a small caloric debt in a child are enormous given the ratio of energy distribution between brain and body."

10. Bogin, "Evolutionary Hypotheses for Human Childhood," 81.

11. See Gould, *Ontogeny and Phylogeny*, 290–94.

12. In traditional hunter-gatherer and horticultural societies, studies have found that even without the advantages of modern medicine and sanitation, people manage to raise about 50 percent of their children to adulthood. Monkeys and apes have a success rate between 14 and 36 percent. That means out of every hundred infants born, humans raise at least fourteen more successfully. Over evolutionary time that has made an enormous difference. Even in protected reserves, chimpanzees and gorillas are essentially at zero population growth and their worldwide numbers are dropping. Humans, however, have grown from small clans numbering in the thousands two hundred thousand years ago to seven billion people that live in every conceivable earthly environment, with more coming all the time. The evolutionary "strategy" of a long, if dangerous, human childhood has clearly succeeded, at least for us, for now.

3: LEARNING MACHINES

1. For more detail see http://users.ecs.soton.ac.uk/harnad/Papers/Py104/pinker.langacq.html—Steven Pinker's exploration of human language and its evolution.

2. Planaria can pass along their personal experience to other flatworms in the

oddest way. Untrained flatworms that eat the ground-up brains of other planaria that have been trained to perform specific tasks will quickly exhibit the same knowledge that the dead flatworms acquired in life. R. Joseph, *The Naked Neuron*, 15.

3. Only four weeks into gestation the first brain cells begin to form at the astonishing rate of 250,000 every minute. Billions of neurons will forge links with billions of other neurons, and eventually trillions upon trillions of connections will be made between cells.

4. Research over the past ten years has illustrated exactly how cognitive, emotional, and social capacities are physically connected to behaviors that can affect us throughout our entire lives. Toxic stress damages developing brain architecture, which can lead to lifelong problems in learning, behavior, and physical and mental health. Scientists now know that chronic, unrelenting stress in early childhood, caused by extreme poverty, repeated abuse, or severe maternal depression, for example, can be toxic to the developing brain. On the other hand, so-called positive stress (moderate, short-lived physiological response to uncomfortable experience) is important and necessary to healthy development. Without the buffering protection of adult support, toxic stress can be built into the body largely through epigenetic processes. For more information, see "The Science of Early Childhood Development" and the Working Paper series from the National Scientific Council on the Developing Child.

5. www.developingchild.harvard.edu/content/publications.html. For more detailed information, see the bibliography, National Scientific Council on the Developing Child, "Children's Emotional Development is Built into the Architecture of Their Brains," 2006.

6. For more on the debate and latest information on exactly how much DNA we have in common with chimpanzees visit http://news.national geographic.com/news/2002/09/0924_020924_dnachimp_2.html.

7. Scientists have found that neurons again overproliferate at a second time in our lives, just before puberty, in a way that they do in the first thirty-six months of life. However, the activity takes place in the prefrontal cortex, not the entire brain. It is almost as if the evolution of the prefrontal cortex required a kind of "second childhood." Connections made during this time that aren't used over the long term are also eventually pruned back.

8. See "A Comparison of Atropine and Patching Treatments for Moderate Amblyopia by Patient Age, Cause of Amblyopia, Depth of Amblyopia, and Other Factors," *Ophthalmology* 110 (8) (August 2003): 1632–37; discussion, 1637–38, and L. Kiorpes and J. A. Movshon "Amblyopia: A Developmental Disorder of the Central Visual Pathways," *Cold Spring Harbor Symposia on Quantitative Biology* 61:39–48, for more information about blindness, the visual cortex, and amblyopia.

4: TANGLED WEBS—THE MORAL PRIMATE

1. The same question struck an anthropologist at the University of Southern California, Christopher Boehm, several years ago, so he surveyed fifty previously completed studies of small, nonliterate tribes and bands that live around the world. He wondered if the way these primal communities handled the complexities of ethics, fair play, and morality might offer some insight into the basics of those behaviors in the rest of us. The popular view of nonliterate societies is that they are more prone to violence or war, but Boehm's research revealed they almost always developed, independently of one another, an egalitarian approach to life; one in which they struggled conscientiously to weigh self-interest and common interest. For example, if a bully acted like a silverback, alpha-male gorilla and attempted to dominate the group, the group responded by shaming, ostracizing, or, in extreme cases, killing the perpetrator to ensure individual rights were protected.

2. For a summary of Dunbar's theories, see pages 122–23 of *Thumbs, Toes, and Tears: And Other Traits That Make Us Human.*

3. For more on this case, see Valerie E. Stone et al., "Selective Impairment of Reasoning About Social Exchange in a Patient with Bilateral Limbic System Damage," *Proceedings of the National Academy of Sciences of the United States of America* 99.17 (2002): 11531–36.

4. It is probably not a coincidence that the neuroimaging studies of people who suffer from different forms of autism find that activity in the orbitofrontal cortex, superior temporal sulcus (STS), and the amygdala is low or nonexistent compared with others who don't have autism. As the experience of R.M. illustrated, these areas of the brain are crucial to social interactions most of us take for granted. Autistics are missing many of the functioning brain structures that allow them to "read" minds. Autistics struggle with grasping the intentions of others, or even comprehending that others have states of mind different from theirs. Depending on how severe the autism is, empathy, sympathy, deception, even joking, are out of the question because they all require seeing life, however briefly, from a point of view other than one's own. This means that the scenario building that comes so naturally to most of us is hard for them. Though scientists don't yet understand why, these newer and more ancient parts of the brain seemingly have been shut down or struggle to communicate with one another.

5: THE EVERYWHERE APE

1. Curtis W. Marean, "When the Sea Saved Humanity," *Scientific American* 303.2 (2010): 54–61.

2. Explore National Geographic's fascinating Genographic Project for details of our past migrations at https://genographic.nationalgeographic.com /genographic/lan/en/atlas.html. There is not universal agreement on these

conclusions, but the information nevertheless provides fascinating insights into how we evolved and came to spread across an entire planet. Another excellent site to visit is http://www.bradshawfoundation.com/journey/.

3. The most recent matrilineal common ancestor shared by all living human beings, also known as Mitochondrial Eve, lived roughly 120–150 millennia ago around East Africa. This is about the same time as *Homo sapiens idaltu*. A study of African genetic diversity headed by Dr. Sarah Tishkoff found that Africa's San people express the greatest genetic diversity among the 113 distinct populations sampled in the research, making them one of fourteen "ancestral population clusters."

4. Dramatic climate fluctuations began 356,000 years ago according to researchers at the Smithsonian Institution and continued until about 50,000 years ago due to the elongated orbit of Earth around the sun. During this time, Africa often grew dry and the planet cold. For more see http://humanorigins.si.edu/evidence/human-evolution-timeline-interactive.

5. Read M. Lozano et al., "Right-Handedness of *Homo heidelbergensis* from Sima De Los Huesos (Atapureca, Spain) 500,000 Years Ago," *Evolution and Human Behavior* 30.5 (2009): 369–76, and http://www.newscientist.com/article/dn17184-ancient-teeth-hint-that-righthandedness-is-nothing-new.html for more details on handedness and brain lateralization at this point in human evolution.

6. See: http://humanorigins.si.edu/evidence/genetics/ancient-dna-and-neanderthals for more insights into Neanderthal range.

7. See http://news.bbc.co.uk/2/hi/science/nature/3948165.stm.

8. For more details on this fascinating theory read "Genetic Analysis of Lice Supports Direct Contact Between Modern and Archaic Humans" at http://www.plosbiology.org/article/info:doi/10.1371/journal.pbio.0020340.

6: COUSIN CREATURES

1. Neanderthal skulls were first discovered in Engis, Belgium (1829), by Philippe-Charles Schmerling and in Forbes' Quarry, Gibraltar (1848), both prior to the specimen discovered in the Neander Valley in Erkrath near Düsseldorf in August 1856. At the time, no one was quite sure what they were. Later they were identified as Neanderthals. If they had initially been identified and investigated further, the species might have been named Gibraltarians or Engiseans rather than Neanderthals.

2. Lighter, straighter hair is often a by-product of lighter, fairer skin.

3. Scientists have speculated that one of the reasons it is so difficult to find fossils of *Homo sapiens* from the same period is that they hadn't yet begun to bury their dead even if Neanderthals had.

4. It's difficult to know how many Native Americans lived in the continental United States before the arrival of white men, but it couldn't have exceeded many more than tens of thousands. In 1823 President James Monroe re-

ported the "Chayenes" to be "a tribe of three thousand two hundred and fifty souls, dwelling and hunting on a river of the same name, a western tributary of the Missouri, a little above the Great Bend." Ten years later, Catlin, the famous painter of Native Americans, reported, "The Shiennes are a small tribe of about three thousand in number, living neighbors to the Sioux on the west of them, between the Black Hills and the Rocky Mountains." In 1822 the population of the two divisions of the Sioux was estimated at nearly thirteen thousand.

5. You can listen to the sound of the Neanderthal's *e* at http://www.fau.edu/explore/media/FAU-neanderthal.wav. It's fascinating.

7: BEAUTIES IN THE BEAST

1. Recounted in Darwin's *Descent of Man*, chap. 19.
2. See *Thumbs, Toes, and Tears: And Other Traits That Make Us Human* for a more detailed exploration of why women evolved large breasts and other insights into the attractions between men and women.
3. J. H. Langlois, L. Kalakanis, A. J. Rubenstein, A. Larson, M. Hallam, and M. Smoot, "Maxims or Myths of Beauty? A Meta-analytic and Theoretical Review." *Psychological Bulletin* 126 (2000). 390–423. Also see http://homepage.psy.utexas.edu/homepage/group/langloislab/facialattract.html.
4. *Descent of Man*, chap. 19.
5. Ibid.
6. At all of these sites researchers found piles of seashells. Together with the much-older evidence from the cave at Pinnacle Point, the shells suggest that seafood may have served as a nutritional trigger at a crucial point in human history, providing the fatty acids that modern humans needed to make an already large and intricate brain faster and smarter. Stanford University paleoanthropologist Richard Klein has long argued that a genetic mutation at roughly this point in human history sparked a sudden increase in brainpower, perhaps linked to the onset of speech.
7. E. Bates, with L. Benigni, I. Bretherton, L. Camaioni, and V. Volterra, *The Emergence of Symbols: Cognition and Communication in Infancy.* New York: Academic Press, 1979. Note the term Bates used in the passage, *heterochrony*, which is defined as a developmental change in the timing of events leading to changes in a living thing's size and shape, is often used interchangeably with *neoteny*.
8. For more on the exponential rate of change in evolution of all kinds from the universe to human culture, explore Ray Kurzweil's concept of the Law of Accelerating Returns, defined in his book *The Age of Spiritual Machines: When Computers Exceed Human Intelligence.*
9. These mutations may also have kicked in the ultimate symbolic ability and the most extreme proof that the human brain had evolved to a point where its owners had become self-aware—modern, human language and speech.

8: THE VOICE INSIDE YOUR HEAD

1. See Belinda R. Lennox, S. Bert, G. Park, Peter B. Jones, and Peter G. Morris, "Spatial and Temporal Mapping of Neural Activity Associated with Auditory Hallucinations," *Lancet* 353 (February 2, 1999), http://www.bmu.psychiatry.cam.ac.uk/PUBLICATION_STORE/lennox99spa.pdf.

2. This story was related in comments online following an article in *Scientific American* entitled "It's No Delusion: Evolution May Favor Schizophrenia Genes" at http://www.scientificamerican.com/article.cfm?id=evolution-may-favor-schizophrenia-genes.

3. Eighty percent of diagnosed autistics are men based on research in C. J. Newschaffer, L. A. Croen, J. Daniels, et al., "The Epidemiology of Autism Spectrum Disorders," *Annual Review of Public Health* 28 (2007): 235–58. doi:10.1146/annurev.publhealth.28.021406.144007.PMID_17367287.

BIBLIOGRAPHY

Ackerman, Jennifer. "The Downside of Upright." ngm.nationalgeo
graphic.com, July 1, 2006, 1–2. http://ngm.nationalgeographic.com
/2006/07/bipedal-body/ackerman-text.

Akst, Jef. "Ancient Humans More Diverse? " the-scientist.com, 2010, 1–3.
http://classic.the-scientist.com/blog/display/56279/.

Amen-Ra, Nūn. "How Dietary Restriction Catalyzed the Evolution of
the Human Brain: An Exposition of the Nutritional Neurotrophic
Neoteny Theory." *Medical Hypotheses* 69.5 (2007): 1147–53.

"Anthropologist's Studies of Childbirth Bring New Focus on Women in
Evolution." www.sciencedaily.com, February 25, 2009. http://www
.sciencedaily.com/releases/2009/02/090217173043.htm?utm
_source=feedburner&utm_medium=feed&utm_campaign=Feed
%3A+sciencedaily+%28ScienceDaily%3A+Latest+Science+News%29.

Bahn, Paul, consulting ed. *Written in Bones: How Human Remains Unlock
the Secrets of the Dead.* Toronto, Ontario: Quintet Publishing, 2003.

Baker, T. J., and J. Bichsel. "Personality Predictors of Intelligence: Differences Between Young and Cognitively Healthy Older Adults." *Personality and Individual Differences* 41.5 (2006): 861–71.

Banks, William E., Francesco d'Errico, A. Townsend Peterson, Masa
Kageyama, Adriana Sima, and Maria-Fernanda Sánchez-Goñi. "Neanderthal Extinction by Competitive Exclusion." *PLoS ONE* 3 (12)
(2008): e3972. doi:10.1371/journal.pone.0003972.

Bates, E. "Competition, Variation, and Language Learning. Mechanisms
of Language Acquisition." Mechanisms of Language Acquistion. Edited by Brian MacWhinney, 157–93. Hillsdale, NJ: Lawrence Erlbaum
Associates, 1987.

Belmonte, Matthew K., et al. "Autism and Abnormal Development of
Brain Connectivity." *Journal of Neuroscience* 24.42 (2004): 9228–31.

Biederman, I., and E. Vessel. "Perceptual Pleasure and the Brain: A Novel Theory Explains Why the Brain Craves Information and Seeks It Through the Senses." *American Scientist* 94.3 (2006): 247–53.

Bloom, Paul. "The Moral Life of Babies." www.nytimes.com, 2010. http://www.nytimes.com/2010/05/09/magazine/09babies-t.html ?pagewanted=all.

Boehm, Christopher. "Political Primates | Greater Good." greatergood .berkeley.edu, December 1, 2008. http://greatergood.berkeley.edu/ar ticle/item/political_primates/.

Bogin, B. A. "Evolutionary Hypotheses for Human Childhood." *Yearbook of Physical Anthropology* 40 (1997): 63–89.

Bond, Charles F., and Bella M. DePaulo. "Accuracy of Deception Judgments." *Personality and Social Psychology Review* 10.3 (2006): 214–34.

"Brain Network Related to Intelligence Identified." www.sciencedaily .com, September 9, 2007. http://www.sciencedaily.com/releases/2007 /09/070911092117.htm.

Briggs, Adrian W., et al. "Targeted Retrieval and Analysis of Five Neandertal mtDNA Genomes." *Transactions of the IRE Professional Group on Audio* 325.5938 (2009): 318–21.

Brockman, John. "Science of Happiness: A Talk with Daniel Gilbert." www.edge.org, May 22, 2006. http://www.edge.org/3rd_culture/gil bert06/gilbert06_index.html.

Brotherson, S. "Understanding Brain Development in Young Children." *Bright Beginnings* 4 (2005).

Brown, Kyle S., et al. "Fire as an Engineering Tool of Early Modern Humans." *Transactions of the IRE Professional Group on Audio* 325.5942 (2009): 859–62.

Brüne, Martin. "Neoteny, Psychiatric Disorders and the Social Brain: Hypotheses on Heterochrony and the Modularity of the Mind." *Anthropology & Medicine* 7.3 (2000): 301–18.

———. "Schizophrenia: An Evolutionary Enigma?" *Neuroscience and Biobehavioral Reviews* 28.1 (2004): 41–53.

Callaway, Ewen. "Neanderthals Speak Out After 30,000 Years." www .newscientist.com, April 15, 2008. http://www.newscientist.com/article /dn13672-neanderthals-speak-out-after-30000-years.html.

Carroll, Sean B. "Genetics and the Making of *Homo sapiens*." *Nature* 422.6934 (2003): 849–57.

Chick, Garry. "What Is Play For?" Keynote address, Association for the Study of Play, St. Petersburg, FL, February 1998.

Cohen, A. S., et al. "Paleoclimate and Human Evolution Workshop." *Eos, Transactions, American Geophysical Union* 87.16 (2006): 161.

"A Comparison of Atropine and Patching Treatments for Moderate Amblyopia by Patient Age, Cause of Amblyopia, Depth of Amblyopia, and

Other Factors." *Ophthalmology* 110 (8) (August 2003): 1632–37; discussion, 1637–38.

Cosmides, L., H. C. Barrett, and J. Tooby. "Colloquium Paper: Adaptive Specializations, Social Exchange, and the Evolution of Human Intelligence." *Proceedings of the National Academy of Sciences of the United States of America* 107, supplement 2 (2010): 9007–14.

Courchesne, Eric E. "Brain Development in Autism: Early Overgrowth Followed by Premature Arrest of Growth." *Developmental Disabilities Research Reviews* 10.2 (2004): 106–11.

Cowley, Geoffrey. "Biology of Beauty." www.thedailybeast.com/newsweek.html, June 2, 1996, 3. http://www.thedailybeast.com/newsweek/1996/06/02/the-biology-of-beauty.html.

"Daniel Dennett's Theory of Consciousness: The Intentional Stance and Multiple Drafts." http://www.consciousentities.com. Accessed April 6, 2011.

Darwin, Charles. *The Descent of Man and Selection in Relation to Sex*. Norwalk, CT: Heritage Press, 1972.

———. *The Origin of the Species*. Hardback ed. New York: Barnes and Noble, 2008.

Dawkins, Richard. *The Blind Watchmaker: Why the Evidence of Evolution Reveals a Universe Without Design*. Trade paperback ed. New York: W. W. Norton, 2006.

———. *The Selfish Gene*. 30th anniversary ed. New York: Oxford University Press, 2009.

Dawson, Geraldine G., et al. "Defining the Broader Phenotype of Autism: Genetic, Brain, and Behavioral Perspectives." *Development and Psychopathology* 14.3 (2002): 581–611.

Deacon, Terrence. *The Symbolic Species: The Co-Evolution of Language and the Brain*. Trade paperback. New York: W. W. Norton, 1998.

Dean, Brian. "Is Schizophrenia the Price of Human Central Nervous System Complexity?" *Australian and New Zealand Journal of Psychiatry* 43.1 (2009): 13–24.

Dean, C. C., et al. "Growth Processes in Teeth Distinguish Modern Humans from *Homo erectus* and Earlier Hominins." *Nature* 414.6864 (2001): 628–31.

Dean, Christopher. "Growing Up Slowly 160,000 Years Ago." *Proceedings of the National Academy of Sciences of the United States of America* 104.15 (2007): 6093–94.

De Waal, Frans B. M., *Chimpanzee Politics: Power and Sex Among Apes*. 25th anniversary ed. Baltimore: Johns Hopkins University Press, 2007.

———. "Do Humans Alone Feel Your Pain?" chronicle.com, October 26, 2011. http://chronicle.com/article/Do-Humans-Alone-Feel-Your/26238/.

————. "Morality and the Social Instincts: Continuity with the Other Primates." *Tanner Lectures on Human Values*, 2003.

DiCicco-Bloom, Emanuel, et al. "The Developmental Neurobiology of Autism Spectrum Disorder." *Journal of Neuroscience* 26.26 (2006): 6897–6906.

"DNA Evidence Tells of Human Migration." www.sciencedaily.com, February 24, 2010. http://www.sciencedaily.com/releases/2010/02/100222121618.htm.

Doyle-Burr, Nora. "New Human Species Discovered? How China Fossils Could Redefine 'Human,'" *Christian Science Monitor*, 2012.

Dreifus, Claudia. "A Conversation with Philip G. Zimbardo; Finding Hope in Knowing the Universal Capacity for Evil." *New York Times*, April 3, 2007. http://www.nytimes.com/2007/04/03/science/03conv.html.

Dyson, Freeman. *Disturbing the Universe*. New York: Basic Books, 1979.

Eiseley, Loren. *The Immense Journey*. Paperback. New York: Vintage Books, 1977.

————. *The Unexpected Universe*. Trade paperback. New York: Harcourt Brace Jovanovich, 1985.

Enard, Wolfgang, et al. "Molecular Evolution of FOXP2, a Gene Involved in Speech and Language." *Nature* 418.6900 (2002): 869–72.

Ermer, E., et al. "Cheater Detection Mechanism." *Encyclopedia of Social Psychology* (2007): 138–40.

Fabre, Virginie V., Silvana S. Condemi, and Anna A. Degioanni. "Genetic Evidence of Geographical Groups Among Neanderthals." *Transactions of the IRE Professional Group on Audio* 4 (4) (January 1, 2009): e5151. doi:10.1371/journal.pone.0005151.

Fagan, Brian. *Cro-Magnon: How the Ice Age Gave Birth to the First Modern Humans*. New York: Bloomsbury Press, 2010.

Fagan, J. F., III. "New Evidence for the Prediction of Intelligence from Infancy." *Infant Mental Health Journal* 3.4 (1982): 219–28.

Falk, Dean. "New Information About Albert Einstein's Brain." www.frontiersin.org/evolutionary_neuroscience 1 (2009): 3. http://www.frontiersin.org/evolutionary_neuroscience/10.3389/neuro.18.003.2009/abstract.

————. "Prelinguistic Evolution in Early Hominins: Whence Motherese?" *Behavioral and Brain Sciences* 27.4 (2004): 491–503.

"Fossil from Last Common Ancestor of Neanderthals and Humans Found in Europe, 1.2 Million Years Old." *Science Daily*, April 4, 2008. Accessed March 17, 2011.

Frankfurt, Harry G. *On Bullshit*. Princeton, NJ: Princeton University Press, 2005.

Friedman, Danielle. "Parent Like a Caveman." www.thedailybeast.com, October 10, 2010. http://www.thedailybeast.com/articles/2010/10/11/hunter-gatherer-parents-better-than-todays-moms-and-dads.html.

Fu, X., et al. "Rapid Metabolic Evolution in Human Prefrontal Cortex." *Proceedings of the National Academy of Sciences of the United States of America* 108.15 (2011): 6181–86.

Furnham, Adrian, and Emma Reeves. "The Relative Influence of Facial Neoteny and Waist-to-Hip Ratio on Judgements of Female Attractiveness and Fecundity." *Psychology, Health & Medicine* 11.2 (2006): 129–41.

Genographic Project. National Geographic Society. https://genographic .nationalgeographic.com/genographic/lan/en/atlas.html.

Ghose, Tia. "Bugs Hold Clues to Human Origins." the-scientist.com, January 22, 2009. http://classic.the-scientist.com/blog/display/55350/. Accessed March 3, 2011.

Godfrey, L. R., and M. R. Sutherland. "Paradox of Peramorphic Paedomorphosis: Heterochrony and Human Evolution." *American Journal of Physical Anthropology* 99.1 (1996): 17–42.

Golovanova, Liubov Vitaliena, et al. "Significance of Ecological Factors in the Middle to Upper Paleolithic Transition." *Current Anthropology* 51.5 (2010): 655–91.

Gopnik, A. "How Babies Think." *Scientific American* 303.1 (2010): 76–81.

Gopnik, A., et al. "Causal Learning Mechanisms in Very Young Children: Two-, Three-, and Four-Year-Olds Infer Causal Relations from Patterns of Variation and Covariation." *Developmental Psychology* 37.5 (2001): 620–29.

Gould, Stephen Jay. *Ontogeny and Phylogeny.* Cambridge, MA: Harvard University Press, 1977.

———. *The Panda's Thumb: More Reflections in Natural History.* Trade paperback. New York: W. W. Norton, 1992.

Grafton, Scott, et al. "Brain Scans Go Legal." *Scientific American*, November 29, 2006, 84.

Grant, Richard P. "Creative Madness." *Scientist* 24.8 (2010): 23–25.

Green, Richard E., et al. "A Draft Sequence of the Neanderthal Genome." *Science* 328.5979 (2010): 710–22.

Greenwood, Veronique. "Truth or Lies: A New Study Raises the Question of Whether Being Honest Is a Conscious Decision at All." seed magazine.com, August 17, 2009. http://seedmagazine.com/content /article/truth_or_lies/.

Griskevicius, Vladas, et al. "Blatant Benevolence and Conspicuous Consumption: When Romantic Motives Elicit Strategic Costly Signals." *Journal of Personality and Social Psychology* 93.1 (2007): 85–102.

Gugliotta, Guy. "The Great Human Migration." www.smithsonianmag .com, July 2008, 1–5. http://www.smithsonianmag.com/history -archaeology/human-migration.html.

Gunz, P., F. L. Bookstein, et al. "Early Modern Human Diversity Suggests Subdivided Population Structure and a Complex Out-of-Africa

Scenario." *Proceedings of the National Academy of Sciences* 106.15 (2009): 6094.

Gunz, Philipp, Simon Neubauer, Bruno Maureille, and Jean-Jacques Hublin. "Brain Development After Birth Differs Between Neanderthals and Modern Humans." *Current Biology* 20.21 (2010): R921–22.

———. "Enlarged Image: Brain Development After Birth Differs Between Neanderthals and Modern Humans" (supplement to the reference above). *Current Biology* 20.21 (November 9, 2010): R921–22. doi:10.1016/j.cub.2010.10.018.

Hadhazy, A. "Think Twice: How the Gut's 'Second Brain' Influences Mood and Well-Being." *Scientific American*, 2010. http://www.scientific american.com/article.cfm.

Haidt, Jonathan. "The New Synthesis in Moral Psychology." *Science* 316.5827 (2007): 998–1002.

Harcourt, Alexander H., and Kelly J. Stewart. *Gorilla Society: Conflict, Compromise and Cooperation Between the Sexes*. Chicago: University of Chicago Press, 2007.

Hattori, Kanetoshi. "Two Origins of Language Evolution: Unilateral Gestural Language and Bilateral Vocal Language, Hypotheses from IQ Test Data." *Mankind Quarterly* 39.4 (1999): 399–436.

Hauser, M., et al. "A Dissociation Between Moral Judgments and Justifications." *Mind & Language* 22.1 (2007): 1–21.

Hazlett, Heather Cody, et al. "Magnetic Resonance Imaging and Head Circumference Study of Brain Size in Autism: Birth Through Age 2 Years." *Archives of General Psychiatry* 62.12 (2005): 1366–76.

Henshilwood, Christopher S., et al. "A 100,000-Year-Old Ochre-Processing Workshop at Blombos Cave, South Africa." *Science* 334.6053 (2011): 219–22.

Hill, Jason, et al. "Similar Patterns of Cortical Expansion During Human Development and Evolution." *Proceedings of the National Academy of Sciences of the United States of America* 107.29 (2010): 13135–40.

Hofstadter, Douglas R. *Godel, Escher, Bach: An Eternal Golden Braid*. 20th anniversary ed. New York: Basic Books, 1999.

"How Long Is a Child a Child? Human Developmental Patterns Emerged More Than 160,000 Years Ago." www.sciencedaily.com, March 14, 2007. http://www.sciencedaily.com/releases/2007/03/070313110614.htm.

Hubel, D. H., T. N. Wiesel. "Binocular Interaction in Striate Cortex of Kittens Reared with Artificial Squint." *Journal of Neurophysiology* (London) 28 (1965): 1041–59.

———. "Receptive Fields and Functional Architecture of Monkey Striate Cortex." *Journal of Physiolog* (London) 195 (1968): 215–43.

———. "Receptive Fields, Binocular Interaction, and Functional Architecture in the Cat's Visual Cortex." *Journal of Physiology* (London) 160 (1962): 106–54.

Irvine, William B. *On Desire: Why We Want What We Want.* New York: Oxford University Press, 2006.

Jaynes, Julian. *The Origin of Consciousness in the Breakdown of the Bicameral Mind.* Boston: Houghton Mifflin, 1976

Joseph, R. *The Naked Neuron.* New York: Plenum Press, 1993.

Jung, Carl C., ed. *Man and His Symbols.* New York: Anchor Books, 1964.

Kelley, Jay, and Gary T. Schwartz. "Dental Development and Life History in Living African and Asian Apes." *Proceedings of the National Academy of Sciences of the United States of America* 107.3 (2010): 1035–40.

"Key Brain Regulatory Gene Shows Evolution in Humans." www.sciencedaily.com, December 12, 2005. http://www.sciencedaily.com/releases/2005/12/051212120211.htm.

Kiorpes L., and J. A. Movshon. "Amblyopia: A Developmental Disorder of the Central Visual Pathways." *Cold Spring Harbor Symposia on Quantitative Biology* 61:39–48.

———. "Behavioral Analysis of Visual Development." In *Development of Sensory Systems in Mammals,* edited by J. R. Coleman, 125–54. New York: Wiley, 1990.

Kiorpes, Lynne, Daniel C. Kiper, Lawrence P. O'Keefe, James R. Cavanaugh, and J. Anthony Movshon. "Neuronal Correlates of Amblyopia in the Visual Cortex of Macaque Monkeys with Experimental Strabismus and Anisometropia." *Journal of Neuroscience* 18 (16) (August 15, 1998): 6411–24.

Konner, Melvin. *The Evolution of Childhood.* Cambridge, MA: Belknap Press of the Harvard University Press, 2010.

Krasnow, Max M., et al. "Cognitive Adaptations for Gathering-Related Navigation in Humans." *Evolution and Human Behavior* 32.1 (2011): 1–12.

Krause, Johannes J., et al. "The Derived FOXP2 Variant of Modern Humans Was Shared with Neanderthals." *Current Biology* 17.21 (2007): 1908–12.

Kubicek, Stefan. "Infographic: Epigenetics—a Primer." *Scientist* 25.3 (2001): 32.

Kurtén, Björn, *Dance of the Tiger: A Novel of the Ice Age.* 3rd ed. New York: Berkeley Books, 1982.

Lambert, David, and the Diagram Group. *The Field Guide to Early Man.* New York: Facts on File, 1987.

Langlois, J. H., L. Kalakanis, A. J. Rubenstein, A. Larson, M. Hallam, and M. Smoot. "Maxims or Myths of Beauty? A Meta-analytic and Theoretical Review." *Psychological Bulletin* 126 (2000): 390–423.

Langlois, Judith. "The Question of Beauty." beautymatters.blogspot.com, February 4, 2000. Accessed April 1, 2011.

"Last Humans on Earth Survived in Ice Age Sheltering Garden of Eden, Claim Scientists." *Daily Mail.* July 27, 2010. http://www.dailymail.co.uk/sciencetech/article-1297765/.html.

Lennox, Belinda R., S. Bert, G. Park, Peter B. Jones, and Peter G. Morris. "Spatial and Temporal Mapping of Neural Activity Associated with Auditory Hallucinations." *Lancet* 353 (February 2, 1999).

Leonard, W. R., and M. L. Robertson. "Evolutionary Perspectives on Human Nutrition: The Influence of Brain and Body Size on Diet and Metabolism." *American Journal of Human Biology* 4 (1992): 179–95.

Leslie, Mitchell. "Suddenly Smarter." *Stanford Magazine*, July 1, 2002, 1–11.

Leutwyler, Kristin. "First Gene for Schizophrenia Discovered." *Scientific American*, March 20, 2001.

Lieberman, Philip P. "On the Nature and Evolution of the Neural Bases of Human Language." *American Journal of Physical Anthropology*, supplement 35 (2002): 36–62.

"Long Legs Are More Efficient, According to New Math Model." www.sciencedaily.com, March 19, 2007, 1–2. http://www.sciencedaily.com/releases/2007/03/070312091455.htm.

Lozano, M., et al. "Right-Handedness of *Homo heidelbergensis* from Sima De Los Huesos (Atapureca, Spain) 500,000 Years Ago." *Evolution and Human Behavior* 30.5 (2009): 369–76.

Maestripieri, Dario. *Machiavellian Intelligence: How Rhesus Macaques and Humans Have Conquered the World.* Chicago: University of Chicago Press, 2007.

Manica, Andrea, et al. "The Effect of Ancient Population Bottlenecks on Human Phenotypic Variation." *Nature* 448.7151 (2007): 346–48.

"Man's Earliest Direct Ancestors Looked More Apelike Than Previously Believed." www.sciencedaily.com, March 27, 2007, 1–2. http://www.sciencedaily.com/releases/2007/03/070324133018.htm. Accessed August 20, 2010.

Marean, Curtis W. "When the Sea Saved Humanity." *Scientific American* 303.2 (2010): 54–61.

Miller, Earl, and Jonathan Cohen. "An Integrative Theory of Prefrontal Cortex Function." *Annual Review of Neuroscience* 24 (2001).

Miller, Geoffrey. *The Mating Mind: How Sexual Choice Shaped the Evolution of Human Nature.* New York: Anchor Books, 2001.

Mithen, Steven. *The Singing Neanderthals: The Origins of Music, Language, Mind and Body.* Cambridge, MA: Harvard University Press, 2006.

"Modern Humans, Arrival in South Asia May Have Led to Demise of Indigenous Populations." www.sciencedaily.com, November 7, 2005. http://www.sciencedaily.com/releases/2005/11/051107080321.htm.

"Modern Man Found to Be Generally Monogamous, Moderately Polygamous." www.sciencedaily.com, March 3, 2010. http://www.sciencedaily.com/releases/2010/03/100302112018.htm.

Morris, Desmond. *The Naked Ape.* First American ed. 3rd printing. New York: McGraw-Hill, 1967.

Murray, Elisabeth A. "The Amygdala, Reward and Emotion." *Trends in Cognitive Sciences* 11.11 (2007): 489–97.

National Scientific Council on the Developing Child. "Young Children Develop in an Environment of Relationships: Working Paper No. 1." 2004, 1–12.

———. "Children's Emotional Development Is Built into the Architecture of Their Brains: Working Paper No. 2." 2006, 1–16.

———. "Early Exposure to Toxic Substances Damages Brain Architecture: Working Paper No. 4." 2006, 1–20.

———. "The Timing and Quality of Early Experiences Combine to Shape Brain Architecture: Working Paper No. 5." 2008, 1–12.

———. "Early Experiences Can Alter Gene Expression and Affect Long-Term Development: Working Paper No. 10." 2010, 1–12.

"Neanderthal Children Grew Up Fast." www.sciencedaily.com, December 5, 2007. http://www.sciencedaily.com/releases/2007/12/071204100409.htm.

"Neanderthals Speak Again After 30,000 Years." www.sciencedaily.com, April 21, 2008. http://www.sciencedaily.com/releases/2008/04/080421154426.htm.

Neill, David. "Cortical Evolution and Human Behavior." *Brain Research Bulletin* 74 (2007): 191–205.

Nettle, Daniel, and Helen Clegg. "Schizotypy, Creativity and Mating Success in Humans." *Proceedings of the Royal Society B* 273.1586 (2006): 611–15.

"New Kenyan Fossils Challenge Established Views on Early Evolution of Our Genus Homo." www.sciencedaily.com, August 13, 2007. http://www.sciencedaily.com/releases/2007/08/070813093132.htm.

Newschaffer, C. J., L. A. Croen, J. Daniels, et al. "The Epidemiology of Autism Spectrum Disorders." *Annual Review of Public Health* 28 (2007): 235–58. doi:10.1146/annurev.publhealth.28.021406.144007.PMID_17367287.

Nieder, Andreas. "Prefrontal Cortex and the Evolution of Symbolic Reference." *Current Opinion in Neurobiology* 19.1 (2009): 99–108.

NIMH. "Teenage Brain: A Work in Progress" (fact sheet). wwwapps.nimh.nih.gov/index.shtml, July 18, 2011. http://wwwapps.nimh.nih.gov/health/publications/teenage-brain-a-work-in-progress.shtml.

Oakley, Barbara. "What a Tangled Web We Weave." the-scientist.com, April 10, 2009, 3. http://classic.the-scientist.com/news/display/55610/.

Olivieri, Anna, et al. "The mtDNA Legacy of the Levantine Early Upper Paleolithic in Africa." *Science* 314.5806 (2006): 1767–70.

Pacchioli, David. "Moral Brain." *Research, University of Pennsylvania* (2006): 5.

Patel, Aniruddh D. *Music, Language, and the Brain*. New York: Oxford University Press, 2008.

Paus, T., et al. "Structural Maturation of Neural Pathways in Children and Adolescents: In Vivo Study." *Science* 283 (March 19, 1999): 1908.

Penin, Xavier, Christine Berge, and Michel Baylac. "Ontogenetic Study of the Skull in Modern Humans and the Common Chimpanzees: Neotenic Hypothesis Reconsidered with a Tridimensional Procrustes Analysis." *American Journal of Physical Anthropology* 118.1 (2002): 50–62.

Perrett, D. I., K. J. Lee, I. Penton-Voak, D. Rowland, S. Yoshikawa, D. M. Burt, S. P. Henzi, D. L. Castles, and S. Akamatsu. "Effects of Sexual Dimorphism on Facial Attractiveness." *Nature* 394.6696 (1998): 884–87.

Pontzer, Herman H. "Predicting the Energy Cost of Terrestrial Locomotion: A Test of the LiMb Model in Humans and Quadrupeds." *Journal of Experimental Biology* 210, pt. 3 (2007): 484–94.

Potts, Richard, and Christopher Solan. *What Does It Mean to Be Human?* Washington, DC: National Geographic, 2010.

Reed, David L., et al. "Genetic Analysis of Lice Supports Direct Contact Between Modern and Archaic Humans." *Transactions of the IRE Professional Group on Audio* 2.11 (2004): e340.

Reich, David D., et al. "Genetic History of an Archaic Hominin Group from Denisova Cave in Siberia." *Nature* 468.7327 (2010): 1053–60.

Riel-Salvatore, Julien. "A Niche Construction Perspective on the Middle–Upper Paleolithic Transition in Italy." *Journal of Archaeological Method and Theory* 17.4 (2010): 323–55.

———. "What Is a 'Transitional' Industry? The Uluzzian of Southern Italy as a Case Study." *Sourcebook of Paleolithic Transitions* (2009): 377–96.

Rightmire, G. Philip. "Human Evolution in the Middle Pleistocene: The Role of *Homo heidelbergensis*." *Evolutionary Anthropology* (2011): 1–10.

Rincon, Paul. "Neanderthals' 'Last Rock Refuge.'" www.bbc.com, September 13, 2006. http://news.bbc.co.uk/2/hi/science/nature/5343266.stm.

———. "Neanderthals 'Not Close Family.'" www.bbc.com, January 27, 2004. http://news.bbc.co.uk/2/hi/science/nature/3431609.stm.

Rosen, Jeffrey. "The Brain on the Stand." *New York Times Magazine*, March 11, 2007, 46–84.

Rosenberg, K. R., and W. R. Trevathan. "The Evolution of Human Birth." *Scientific American* 285.5 (2001): 72–77.

Rozzi, Fernando V. Ramirez, and José Maria Bermudez De Castro. "Surprisingly Rapid Growth in Neanderthals." *Nature* 428.6986 (2004): 936–39.

Sawyer, G. J., and Viktor Deak. *The Last Human*. New Haven, CT: Yale University Press, 2007.

"Schizophrenia: Costly By-Product of Human Brain Evolution?" *Science Daily*, August 5, 2008. http://www.sciencedaily.com/releases/2008/08/080804222910.htm.

Sell, A., J. Tooby, and L. Cosmides. "Formidability and the Logic of Human Anger." *Proceedings of the National Academy of Sciences of the United States of America* 106.35 (2009): 15073–78.

Sell, Aaron A., et al. "Human Adaptations for the Visual Assessment of Strength and Fighting Ability from the Body and Face." *Proceedings of the Royal Society B* 276.1656 (2009): 575–84.

"Sign In to Read: Neanderthal Body Art Hints at Ancient Language." *New Scientist*, March 29, 2011. http://www.newscientist.com/article/mg19726494.600-neanderthal-body-art-hints-at-ancient-language.html.

Silberman, S. "Don't Even Think About Lying: How Brain Scans Are Reinventing the Science of Lie Detection." *Wired San Francisco* 14.1 (2006): 142.

Sinclair, David A., and Lenny Guarente. "Unlocking the Secrets of Longevity Genes." *Scientific American* 294.3 (2006): 48–51, 54–57.

Singer, Emily. "An Innate Ability to Smell Scams." *Los Angeles Times*, August 19, 2002. http://articles.latimes.com/2002/aug/19/science/sci-cheat19.

Slimak, L., et al. "Late Mousterian Persistence near the Arctic Circle." *Science* 332.6031 (2011): 841–45.

Smith, Tanya M., et al. "Earliest Evidence of Modern Human Life History in North African Early *Homo sapiens*." *Proceedings of the National Academy of Sciences of the United States of America* 104.15 (2007): 6128–33.

Smith, Tanya M., et al. "Dental Evidence for Ontogenetic Differences Between Modern Humans and Neanderthals." *Proceedings of the National Academy of Sciences of the United States of America* 107.49 (2010): 20923–28.

Sockol, Michael D., David A. Raichlen, and Herman H. Pontzer. "Chimpanzee Locomotor Energetics and the Origin of Human Bipedalism." *Proceedings of the National Academy of Sciences of the United States of America* 104.30 (2007): 12265–69.

Sparks, B. F., et al. "Brain Structural Abnormalities in Young Children with Autism Spectrum Disorder." *Neurology* 59.2 (2002): 184–92.

Stone, Valerie E., et al. "Selective Impairment of Reasoning About Social Exchange in a Patient with Bilateral Limbic System Damage." *Proceedings of the National Academy of Sciences of the United States of America* 99.17 (2002): 11531–36.

"Study Identifies Energy Efficiency as Reason for Evolution of Upright Walking." *Science Daily*, July 17, 2007. http://www.sciencedaily.com/releases/2007/07/070716191140.htm.

"Supervolcano Eruption—in Sumatra—Deforested India 73,000 Years Ago." *Science Daily*, November 24, 2009. http://www.sciencedaily.com/releases/2009/11/091123142739.htm.

Swaminathan, Nikhil. "It's No Delusion: Evolution May Favor Schizophrenia Genes." *Scientific American*, September 6, 2007.

————. "White Matter Matters in Schizophrenia." *Scientific American*, April 24, 2011.

Tattersall, I. "Once We Were Not Alone." *Scientific American* 282.1 (2000): 56–62.

Texier, Pierre-Jean, et al. "A Howiesons Poort Tradition of Engraving Ostrich Eggshell Containers Dated to 60,000 Years Ago at Diepkloof Rock Shelter, South Africa." *Proceedings of the National Academy of Sciences of the United States of America* 107.14 (2010): 6180–85.

Thomas, Lewis. *The Medusa and the Snail*. New York: Viking Press, 1979.

"Three Neanderthal Sub-Groups Confirmed." *Science Daily*, April 15, 2009. http://www.sciencedaily.com/releases/2009/04/090415075150 .htm.

"Toba Catastrophe Theory." *Science Daily*, n.d. http://www.sciencedaily. com/articles/t/toba_catastrophe_theory.htm. Accessed March 9, 2011.

Tooby, J., and L. Cosmides. "Groups in Mind: The Coalitional Roots of War and Morality." *Human Morality & Sociality: Evolutionary & Comparative Perspectives*. New York: Palgrave Macmillan, 2010.

Tooby, J., and I. DeVore. "The Reconstruction of Hominid Behavioral Evolution Through Strategic Modeling." In *The Evolution of Human Behavior: Primate Models*, edited by Warren G. Kinsey, 183–237. Albany: State University of New York Press, 1987.

Tzedakis, P. C., K. A. Hughen, I. Cacho, and K. Harvati. "Placing Late Neanderthals in a Climatic Context." *Nature* 449 (7159) (September 13, 2007): 206–8. doi:10.1038/nature06117.

Van Wyhe, John. *The Darwin Experience: The Story of the Man and His Theory of Evolution*. Washington, DC: National Geographic, 2008.

Volk, T., and J. Atkinson. "Is Child Death the Crucible of Human Evolution?" *Journal of Social, Evolutionary and Cultural Psychology* 2 (2008): 247–60.

Vrba, E. S. "Climate, Heterochrony, and Human Evolution." *Journal of Anthropological Research* (1996): 1–28.

Wade, Nicholas. "Scientist Finds the Beginnings of Morality in Primate Behavior." *New York Times*, March 20, 2007. http://www.nytimes.com /2007/03/20/science/20moral.html?_r=1&pagewanted=all.

————. "Signs of Neanderthals Mating with Humans." *New York Times*, May 5, 2007. http://www.nytimes.com/2010/05/07/science/07nean derthal.html.

————. "Tools Suggest Earlier Human Exit from Africa." *New York Times*, January 28, 2011. http://www.nytimes.com/2011/01/28/science /28africa.html?pagewanted=all.

Walter, Chip. *Thumbs, Toes, and Tears: And Other Traits That Make Us Human*. New York: Walker, 2006.

Weaver, Timothy D., and Jean-Jacques Hublin. "Neanderthal Birth Canal Shape and the Evolution of Human Childbirth." *Transactions of the*

IRE Professional Group on Audio 106 (20) (May 19, 2009): 8151–56. doi:10.1073/pnas.0812554106.

Wesson, Kenneth. "Neuroplasticity." *Brain World*, August 26, 2010. http://brainworldmagazine.com/neuroplasticity.

"What Does It Mean to Be Human?" Smithsonian Institution, 2010. http://humanorigins.si.edu.

"Why Humans Walk on Two Legs." *Science Daily*, July 7, 2007. http://www.sciencedaily.com/releases/2007/07/070720111226.htm.

"Why Music?" *Economist*, December 18, 2008, 1–1. http://www.econo mist.com/node/12795510.

"Why We Are, as We Are." *Economist*, December 18, 2008. http://www .economist.com/node/12795581.

Wills, Christopher. *The Runaway Brain: The Evolution of Human Unique-ness*. New York: HarperCollins Publishers, 1993.

Wilson, David Sloan. *Evolution for Everyone: How Darwin's Theory Can Change the Way We Think About Our Lives*. New York: Bantam Dell, 2007.

Wilson, Edward O. *On Human Nature*. Trade paperback. Cambridge, MA: Harvard University Press, 1978.

Wong, K. "Who Were the Neanderthals?" *Scientific American* 289 (2003): 28–37.

Zak, Paul J. "The Neurobiology of Trust." *Scientific American* 298.6 (2008): 88–92, 95.

Zilhão, et al. "Symbolic Use of Marine Shells and Mineral Pigments by Iberian Neanderthals." *Proceedings of the National Academy of Sciences of the United States of America* 107.3 (2010): 1023–28.

Zimmer, Carl. "Siberian Fossils Were Neanderthals' Eastern Cousins, DNA Reveals." *New York Times*, December 23, 2010. http://www.ny times.com/2010/12/23/science/23ancestor.html.

Zipursky, Lawrence S. "Driving Self-Recognition." *American Scientist* 24.11: 40–48.

INDEX

A NOTE ON THE AUTHOR

Chip Walter is founder of the popular website AllThingHuman.net, a former CNN bureau chief, and feature film screenwriter. He has written and produced several award-winning science documentaries for PBS, in collaboration with the National Academy of Sciences, including programs for the Emmy Award–winning series *Planet Earth* and *Infinite Voyage*. Chip's science writing has embraced a broad spectrum of fields and topics. He is the author of *Space Age*, the companion volume to the PBS series of the same title; *I'm Working on That*, written with William Shatner; and *Thumbs, Toes, and Tears—And Other Traits That Make Us Human*. His books have been published in six languages.

Chip's articles have also been featured in the *Economist*, *Scientific American*, *Scientific American Mind*, *Slate*, the *Washington Post*, the *Boston Globe*, *Discover* magazine, and many other publications and websites. He is currently an adjunct professor at Carnegie Mellon University's School of Computer Science and Entertainment Technology Center. He lives in Pittsburgh with his wife, Cyndy, and their children Molly, Steven, Hannah, and Annie.